Lecture Notes in Computer Science   11078

Commenced Publication in 1973
Founding and Former Series Editors:
Gerhard Goos, Juris Hartmanis, and Jan van Leeuwen

More information about this series at http://www.springer.com/series/7409

Claudia d'Amato · Martin Theobald (Eds.)

Reasoning Web

Learning, Uncertainty, Streaming, and Scalability

14th International Summer School 2018
Esch-sur-Alzette, Luxembourg, September 22–26, 2018
Tutorial Lectures

 Springer

Editors
Claudia d'Amato
University of Bari Aldo Moro
Bari
Italy

Martin Theobald
University of Luxembourg
Esch-sur-Alzette
Luxembourg

ISSN 0302-9743   ISSN 1611-3349 (electronic)
Lecture Notes in Computer Science
ISBN 978-3-030-00337-1   ISBN 978-3-030-00338-8 (eBook)
https://doi.org/10.1007/978-3-030-00338-8

Library of Congress Control Number: 2018953706

LNCS Sublibrary: SL3 – Information Systems and Applications, incl. Internet/Web, and HCI

This Springer imprint is published by the registered company Springer Nature Switzerland AG
The registered company address is: Gewerbestrasse 11, 6330 Cham, Switzerland

Preface

The research areas of the Semantic Web, Linked Data and Knowledge Graphs have received a lot of attention in academia and industry recently. Since its inception in 2001, the Semantic Web has aimed at enriching the existing Web with meta-data and processing methods, so as to provide Web-based systems with intelligent capabilities such as context-awareness and decision support. Over the years, the Semantic Web vision has been driving many community efforts which have invested substantial resources in developing vocabularies and ontologies for annotating their resources semantically. Besides ontologies, rules have long been a central part of the Semantic Web framework and are available as one of its fundamental representation tools, with logic serving as a unifying foundation. Linked data is a related research area which studies how one can make RDF data available on the Web and interconnect it with other data with the aim of increasing its value for everybody. Knowledge Graphs have been shown useful not only for Web search (as demonstrated by Google, Bing, etc.) but also in many other application domains.

In 2018, Reasoning Web, the 14th Reasoning Web Summer School, took place in Esch-sur-Alzette at the new Belval Campus of the University of Luxembourg during September 22–26 and was part of the Luxembourg Logic for AI Summit (LuxLogAI: https://luxlogai.uni.lu).

Topics and Lecturers

The program of 2018 included nine lectures covering the following topics:

1. Guido Governatori (CSIRO/Data61)	Practical Normative Reasoning with Defeasible Deontic Logic
2. Hannah Bast (University of Freiburg)	Efficient SPARQL Queries on Very Large Knowledge Graphs
3. Thomas Lukasiewicz and Ismail Ilkan Ceylan (University of Oxford)	A Tutorial on Query Answering and Reasoning over Probabilistic Knowledge Bases
4. Hendrik ter Horst, Matthias Hartung and Philipp Cimiano (Bielefeld University)	Cold-start Knowledge Base Population using Ontology-based Information Extraction with Conditional Random Fields
5. Heiko Pauleim (University of Mannheim)	Machine Learning with and for Knowledge Graphs
6. Daria Stepanova (Max Planck Institute for Informatics, Saarbrücken)	Rule Induction and Reasoning over Knowledge Graphs

(Continued)

7. Steffen Staab and Daniel Janke (University of Koblenz-Landau)	Storing and Querying Semantic Data in the Cloud
8. Emanuele Della Valle (Politecnico di Milano)	Engineering of Web Stream Processing Applications
9. Jacopo Urbani (Vrije Universiteit Amsterdam)	Reasoning at Scale (Tutorial)

Applications and Acknowledgements

We received 55 applications from a broad spectrum of Master and PhD students, Post-Docs, and Professionals in the Semantic Web domain, of which 53 have been accepted for participation at the Summer School and 4 have been selected for a travel grant (including PhD students from Brazil, France, Germany, and Sweden).

We specifically thank the University of Luxembourg and the Luxembourg National Research Fund (FNR) for their support in the organization of the 14th Reasoning Web Summer School in Luxembourg, and we are looking forward to an interesting and exciting program!

July 2018   Martin Theobald
                                                                Claudia d'Amato

Organization

Summit Chair

Leon van der Torre   University of Luxembourg, Luxembourg

General Chair

Xavier Parent   University of Luxembourg, Luxembourg

Program Chairs

Christoph Benzmüller   University of Luxembourg, Luxembourg,
and Free University of Berlin, Germany
Francesco Ricca   University of Calabria, Spain

Proceedings Chair

Dumitru Roman   SINTEF and University of Oslo, Norway

Industry Track Chair

Silvie Spreeuwenberg   LibRT Amsterdam, The Netherlands

Doctoral Consortium Chairs

Kia Teymourian   Boston University, USA
Paul Fodor   Stony Brook University, USA

International Rule Challenge Chairs

Giovanni De Gasperis   University of L'Aquila, Italy
Wolgang Faber   Alpen-Adria-Universität Klagenfurt, Austria
Adrian Giurca   Brandenburgische Technische Universität
Cottbus-Senftenberg, Germany

Reasoning Web (RW) Summer School

Claudia d'Amato   University of Bari, Italy
Martin Theobald   University of Luxembourg, Luxembourg

Sponsorship Chair

TBA

Publicity Chairs

Amal Tawakuli	University of Luxembourg, Luxembourg
Frank Olken	Frank Olken Consulting, USA
Xing Wang	.

Financial Chair

Martin Theobald	University of Luxembourg, Luxembourg

Poster Chair

Alex Steen	Free University of Berlin, Germany

Program Committee

Sudhir Agarwal	Stanford University, USA
Nick Bassiliades	Aristotle University of Thessaloniki, Greece
Christoph Benzmüller (Chair)	Free University of Berlin, Germany
Leopoldo Bertossi	Carleton University, Canada
Mehul Bhatt	University of Bremen, Germany, and Örebro University, Sweden
Pedro Cabalar	University of Corunna, Italy
Diego Calvanese	Free University of Bozen-Bolzano, Italy
Iliano Cervesato	Carnegie Mellon University, USA
Horatiu Cirstea	Loria Nancy, France
Stefania Costantini	Università dell'Aquila, Italy
Juergen Dix	Clausthal University of Technology, Germany
Thomas Eiter	Vienna University of Technology, Austria
Esra Erdem	Sabanci University, Turkey
Wolfgang Faber	Alpen-Adria-Universität Klagenfurt, Austria
Fred Freitas	Universidade Federal de Pernambuco, Brazil
Daniel Gall	Ulm University, Germany
Giancarlo Guizzardi	Federal University of Espirito Santo, Brazil
Michael Kifer	Stony Brook University, USA
Matthias Klusch	Deutsches Forschungszentrum für Künstliche Intelligenz GmbH Saarbrücken, Germany
Michael Kohlhase	Friedrich-Alexander-Universität Erlangen-Nürnberg, Germany
Roman Kontchakov	Birkbeck University of London, UK

Sponsors

Gold Sponsors

Logical Contracts, Binarypark

Contents

Practical Normative Reasoning with Defeasible Deontic Logic

Guido Governatori[(⊠)]

Data61, CSIRO, Dutton Park, Australia
guido.governatori@data61.csiro.au

Abstract. We discuss some essential issues for the formal representation of norms to implement normative reasoning, and we show how to capture those requirements in a computationally oriented formalism, Defeasible Deontic Logic, and we provide the description of this logic, and we illustrate its use to model and reasoning with norms with the help of legal examples.

1 Introduction: Two Normative Reasoning Scenarios

The aim of this contribution is to provide an introduction to a practical computational approach to normative reasoning. To this end we start by proposing two scenarios illustrating some distinctive features of normative reasoning.

Example 1. License for the evaluation of a product [16,26].

Article 1. The Licensor grants the Licensee a license to evaluate the Product.
Article 2. The Licensee must not publish the results of the evaluation of the Product without the approval of the Licensor; the approval must be obtained before the publication. If the Licensee publishes results of the evaluation of the Product without approval from the Licensor, the Licensee has 24 h to remove the material.
Article 3. The Licensee must not publish comments on the evaluation of the Product, unless the Licensee is permitted to publish the results of the evaluation.
Article 4. If the Licensee is commissioned to perform an independent evaluation of the Product, then the Licensee has the obligation to publish the evaluation results.
Article 5. This license terminates automatically if the Licensee breaches this Agreement.

Suppose that the licensee evaluates the product and publishes on her website the results of the evaluation without having received an authorisation from the licensor. Suppose also that the licensee realises that she was not allowed to publish the results of the evaluation, and removes the published results from their website within 24 h from the publication. Is the licensee still able to legally use the product? Since the contract contains a remedial clause (removal within

C. d'Amato and M. Theobald (Eds.): Reasoning Web 2018, LNCS 11078, pp. 1–25, 2018.
https://doi.org/10.1007/978-3-030-00338-8_1

24 h remedies unauthorised publication), it is likely that the license to use the product still holds.

Suppose now, that the licensee, right after publishing the results, posted a tweet about the evaluation of the product and that the tweet counts as commenting on the evaluation. In this case, we have a violation of Article 3, since, even if the results were published, according to Article 2 the publication was not permitted. Thus, she is no longer able to legally use the product under the term of the license.

The final situation we want to analyse is when the publication and the tweet actions take place after the licensee obtained permission for publication. In this case, the licensee has the permission to publish the result and thus they were free to post the tweet. Accordingly, she can continue to use the product under the terms of the license.

Example 2. Consider the following fragment of a contract from the provision of goods and services taken from [10].

3.1 A "Premium Customer" is a customer who has spent more that $10000 in goods.

3.2 Services marked as "special order" are subject to a 5% surcharge. Premium customers are exempt from special order surcharge.

5.2 The (Supplier) shall on receipt of a purchase order for (Services) make them available within one day.

5.3 If for any reason the conditions stated in 4.1 or 4.2 are not met the (Purchaser) is entitled to charge the (Supplier) the rate of $100 for each hour the (Service) is not delivered.

Clause 3.1 provides a definition of Premium Customer for the purpose of the contract. Then Clause 3.2 provides a recipe to compute the price for service marked as special order. In this case, we have two conditions, the standard condition specifying that the price for special order has to be incremented by 5%, and an exception to that computation, when the customer is a Premium Customer. Clause 5.2 establishes an obligation on one of the subject of the norms (a party in the contract, the supplier) based on a condition (that a purchase order had been issued). While we cannot give the complete meaning of Clause 5.3 (it depends on Clauses 4.1 and 4.2, not given here), we can infer that the is triggered in response to a violation of either the conditions specified in 4.1 and 4.2, and that the effect produced is an "entitlement" or in other words a permission.

2   Foundations for Normative Reasoning

The scenarios depicted in the examples illustrate many salient and characteristic aspects of normative reasoning. The first aspect we want to focus on is that when we are given a set of norms normative reasoning is concerned about what are the normative effects that follows from the set of norms, and when such normative effects are in force (or in other terms, when they produce effects that are "legally" binding). As we will elaborate further in the rest of the paper, we will distinguish between two types of "normative" effects. The first type is that some norms (for example Article 3.1 and Article 3.2 in Example 2) specify the meaning of terms/concepts in the context where the norms are valid. The second type of effects determine constraints affecting the subjects of the norms (essentially all other Articles in the two examples above). In both cases, the norms consist of two parts the normative effects, and the conditions under which the normative effects are in force (and when they are in force), for example, taking Article 2 of Example 1 the normative effect is the prohibition of publishing the result of the evaluation of a product, and the condition is not having the authorisation from the licensor to publish. Accordingly, norms can be represented as *if... then... rules*, where the *if* part determines the conditions of applicability of the norm, and the *then* part gives the normative effects.

Based on the discussion so far, we can consider a normative system as a set of clauses (norms), where the clauses/norms are represented as *if... then* rules. Every clause/norm is represented by one (or more) rule(s) with the following form:

$$A_1, \ldots, A_n \hookrightarrow C \tag{1}$$

where $A_1, \ldots, A_n$ and the conditions of applicability of the norm and C is the "effect" of the norm. According to the type of effect we can classify the norms/rules as

- constitutive (also known as counts-as) rules that define the terms used in the normative systems, or in other terms they create "institutional facts" from brute facts and other brute facts.[1]
- prescriptive rules, that determine what "normative" effects are in force based on the conditions of applicability.

For normative effects we consider:

- Obligation,
- Prohibition,
- Permission.

These notions can be defined as follows [30]:

Obligation: a state, an act, or a course of action to which a Bearer is legally bound, and which, if it is not achieved or performed, results in a Violation.

Prohibition: a state, an act, or a course of action to which a Bearer is legally bound, and which, if it is achieved or performed, results in a Violation.

Permission: something is permitted if a Bearer has no Prohibition or Obligation to the contrary. A weak Permission is the absence of the Prohibition or Obligation to the contrary; a strong Permission is an exception or derogation of the Prohibition or Obligation to the contrary.

---

[1] For extended discussions on constitutive or counts-as rules, see the seminal work by Searle [33] and, for formal treatments, the comprehensive [25].

One of the functions of norms is to regulate the behaviour of their subjects by imposing constraints on what the subjects can or cannot do, what situations are deemed legal, and which ones are considered to be illegal. There is an important difference between the constraints imposed by norms and other types of constraints. Typically, a constraint means that the situation described by the constraint cannot occur. For example, suppose you have a constraint A. This means that if $\neg A$ (the negation of A, that is, the opposite of A) occurs, then we have a contradiction, or in other terms, we have an impossible situation. Norms, on the other hand, can be violate $\neg A$, we do not have a contradiction, but rather a violation, or in other terms we have a situation that is classified as "illegal". From a logical point of view, we cannot represent the constraint imposed by a norm simply by A, since the conjunction of A and $\neg A$ is a contradiction. Thus, we need a mechanism to identify the constraints imposed by norms. This mechanism is provided by modal (deontic) operators.

2.1   Deontic Operators

Here we are going to consider the following deontic operators: O for obligation, F for prohibition (forbidden), and P for permission. The deontic operators are modal operators. A modal operator applies to a proposition to create a new proposition where the modal operator qualifies the "truth" of the proposition the operator is applied to. Consider, for instance, based on Example 1, the proposition *publishEvaluation* meaning that "the licensee published the results of the evaluation of the product". We can distinguish the following propositions/statements:

- *publishEvaluation*: this is a factual statement that is true if the licensee has published the evaluation (of the product), and false otherwise (in this case *¬publishEvaluation* is true).
- O*publishEvaluation*: this is a deontic statement meaning that the licensee has the obligation to publish the evaluation of the product. The statement is true, if the obligation to publish the evaluation is in force in the particular case.
- F*publishEvaluation*: this is a deontic statement meaning that the licensee has the prohibition to publish the evaluation of the product. The statement is true, if the prohibition to evaluate the produce is in force in the particular case.
- P*publishEvaluation*: this is a deontic statement meaning that the licensee has the permission to publish the evaluation of the product. The statement can be evaluated as true, if the permission to publish the evaluation is in force in the particular case.

Looking again at Example 1, we have that the prohibition to publish the evaluation of the product is in force (thus, F*publishEvaluation* is true) if the licensee does not get the approval for publication from the licensor, or if she is not commissioned to evaluate the product. However, if she gets the approval (Article 2), then the permission to publish is in force (P*publishEvaluation* holds), and if she is commissioned, then she is under the obligation to publish the evaluation.

Now the question is if she is commissioned to evaluate the product (and thus have the obligation to publish the evaluation), then does she have the permission to publish the evaluation? To provide an answer to this question, we go back to the idea that norms define constraints on what is legal and what is illegal, and how to model a violation. Obligations and prohibitions can be violated; according to some legal scholars, the possibility of being violated can be used to define an obligation (prohibition). A violation means that the content of the obligation has not been met. As we have alluded to above, it is important to notice that a violation does not result in an inconsistency. A violation is, basically, a situation where we have[2]

$$\mathsf{O}A \text{ and } \neg A, \text{ or } \mathsf{F}A \text{ and } A. \tag{2}$$

On the other hand, a permission, as we have already said, is the lack of the obligation to the contrary (and a violation requires an obligation it violates), thus permissions cannot be violated. Now, going back to the question, suppose that we have the obligation to publish the evaluation but we do not have the corresponding permission; this means that we have the prohibition of publishing the evaluation. Thus, we have at the same time the obligation and the prohibition of publishing the evaluation. We have two cases, we publish the evaluation, complying with the obligation to publish, but we violated the prohibition; we refrain from publishing the evaluation complying with prohibition (lack of permission), this violates the obligation to publish. In both cases, we get a violation, while, the intuition is that if we have the obligation to publish, and we do publish, we are fully compliant with the norms. Thus, we can conclude that, in case, we have an obligation, we should have (or derive) the corresponding permission.

From the definitions of the normative effects and the discussion above, we get the following equivalences:

$$\mathsf{F}A \equiv \mathsf{O}\neg A \qquad \mathsf{O}A \equiv \mathsf{F}\neg A \qquad \mathsf{P}A \equiv \neg\mathsf{O}\neg A \tag{3}$$

The first two equivalences state that a prohibition corresponds to a negative obligation, and that a permission is the lack of the obligation to the contrary, or more precisely, that the obligation to the contrary is not in force. However, for permission, we can distinguish two notions: *strong permission*, which specifies that the obligation to the contrary is not in force because it has been derogated by the explicit permission; and *weak permission* simply meaning that the obligation to the contrary is not in force (this can be because the conditions of applicability for the obligation to the contrary do not hold, or simply because there are not norms for it). We will use $\mathsf{P_w}$ to indicate weak permission, $\mathsf{P_s}$ for strong permission and P where the strength of the permission is not specified. We further assume that, for every proposition A, $\mathsf{O}A \rightarrow \mathsf{P_s}A$,[3] and $\mathsf{P_s}A \rightarrow \mathsf{P_w}A$

---

[2] In (modal/deontic) logics where $\mathsf{O}A$ does not imply A a violation does not imply a contradiction, thus there are consistent state to represent situations corresponding to violations.

[3] In this section we use "$\rightarrow$" to indicate material implication of classical propositional logic. In the rest of the paper it will be used to denote a particular type of rules in Defeasible Logic.

and that the equivalence $\neg O \neg A \equiv \mathsf{P_w} A$ for weak permission but not for strong permission[4].

2.2 Handling Violations

Given that a violation is not a contradiction, but a situation where we have OA and $\neg A$, then we can reason about violations, and having provisions depending on violations. An important issue is related to the so called *contrary-to-duty obligations* (CTD)[5]. Shortly a contrary-to-duty obligation is an obligation (or more generally a normative effect) that enters in force because another obligation has been violated. The general structure of a CTD is

1. an obligation OA, and
2. a norm such that the opposite of the content of the obligation is part of the conditions of applicability of the norm, so something with the following structure $C_1, \ldots, C_n, \neg A \hookrightarrow OB$

When the obligation OA is in force, and the conditions of applicability of the norm holds, we have (1) a violation, and (2) the obligation OB enters in force. However, if $C_1, \ldots, C_n$ do no contain OA, then the two provisions are somehow independent, and the CTD is somehow accidental. Nevertheless, normative systems, typically contain prescriptions related to violations, to establish penalties to compensate for the violation (an instance of this is Article 2 in Example 1); or, in some other cases, they contain provisions for ancillary penalties or normative effects that, per se, do not compensate for the violation (this is the case of clause 5.3 in Example 2). In both cases they should make explicit the relationship with a violation. Accordingly, following [14,21] we introduce the operator ($\otimes$) proposed in [21] for CTD and reserve it for *compensatory obligations*, and we distinguish between norms

$$C_1, \ldots, C_n, OA, \neg A \hookrightarrow OB \tag{4}$$

$$C_1, \ldots, C_n \hookrightarrow A \otimes B \tag{5}$$

A norm with the form as in 4 means that the obligation of B requires to have a violation of A as its conditions to be in force. While $A \otimes B$ means that OA is in force, but if violated (i.e., $\neg A$ holds) then OB is in force, and the fulfilling the obligation of B compensates for the violation of the obligation of A.

Notice that not all violations are compensable, this means that there are situations, that while logically consistent, are not legal according to a normative system, and normative systems can contain provisions about this type of situations, see the termination clause, Article 5 in Example 1. An instance of such a case, is when the licensee is commissioned to evaluate the produce, but she

---

[4] Suppose that you do not have an explicit permission for $\neg A$; this means that there are no rules to derogate the obligation of A, but it does not mean that there are no norms mandating A.

[5] The term contrary-to-duty was coined by Chisholm [6]; his work inspired a wealth of research on deontic logic to address the paradox proposed in [6] and other CTD paradoxes. For overviews of such paradoxes and development in the field see [5,31].

does not (Article 4). Another instance, is when the licensee publishes the results of the evaluation without approval, and fails to remove within the allotted time (Article 2).

2.3   Defeasibility: Handling Exceptions

So far we have assumed that norms are represented by normative conditionals with the form $A_1, \ldots, A_n \hookrightarrow C$, but we have not discussed the nature of such conditionals, which are generally defeasible and do not correspond to the if-then material implication of classical propositional logic. Norms are meant to provide general principles, but at the same time they can express exceptions to the principle. It is well understood in Legal Theory and Artificial Intelligence and Law [8,32] that, typically, there are different types of "normative conditionals", but in general normative conditionals are defeasible. Defeasibility is the property that a conclusion is open in principle to revision in case more evidence to the contrary is provided, the conclusion is obtained using only the information available when the conclusion is derived. Defeasible reasoning is in contrast to monotonic reasoning of propositional logic, where no revision is possible. In addition, defeasible reasoning allows reasoning in the face of contradictions, which gives rise to ex false quodlibet in propositional logic. One application of defeasible reasoning is the ability to model exceptions in a simple and natural way.

The first use of defeasible rules is to capture conflicting rules/norms without making the resulting set of rules inconsistent. The following two rules conclude with the negation of each other

$$A_1, \ldots, A_n \hookrightarrow C \tag{6}$$

$$B_1, \ldots, B_m \hookrightarrow \neg C \tag{7}$$

without defeasibile rules, rules with conclusions that are negations of each other could give rise, should $A_1, \ldots, A_n$ and $B_1, \ldots, B_m$ both hold, to a contradiction, i.e., C and $\neg C$, and consequently *ex falso quodlibet*. Instead, defeasible reasoning is sceptical; that is, in case of a conflict such as the above, it refrains from taking any of the two conclusions, unless there are mechanisms to solve the conflict.

Norms, typically give the general conditions of applicability to guarantee that some particular conclusions hold, and then list the possible exceptions (Articles 2, 3 in Example 1 and Clause 3.2 in Example 2 provide explicit instances of this phenomenon, while the exception is implicit in Article 1, it assumes an implicit norm that forbids the use of the product without a license). Exceptions limit the applicability of basic norms/rules. The general structure to model exceptions is given by the following rules:

$$A_1, \ldots, A_n \hookrightarrow C \tag{8}$$

$$A_1, \ldots, A_n, E_1 \hookrightarrow \neg C \tag{9}$$

$$\ldots$$

$$A_1, \ldots, A_n, E_m \hookrightarrow \neg C \tag{10}$$

In this case, the rules from is more specific than the first, and thus it forms an exception to the first, i.e., a case where the rule has extra conditions. In a classical logic setting this would result in contradiction as soon as the conditions of applicability of the general norm/rule (8) and the any of the exceptions (E_i) holds. In defeasible reasoning the extra condition encodes the exception, and blocking the conclusion of the first rule. It is possible to use classical (monotonic) reasoning by encoding (8) as

$$A_1 \wedge \cdots \wedge A_n \wedge \bigwedge_{i=1}^{m} \neg E_i \rightarrow C \tag{11}$$

The above representation still suffers from several drawbacks (for a detailed discussion see [20]). To apply this rule, we have to know for each E_i whether it is true or false. This means that if in a (legal) case we only have partial knowledge of the facts of the cases we cannot draw conclusions, and in the extreme case this approach would require factually omniscient knowledge. The second drawback depends on the possibility of having exceptions to the exceptions, thus for each exception E_i, we could have rules

$$A_1, \ldots, A_n, E_i, F_1^i \hookrightarrow C \tag{12}$$

$$\ldots$$

$$A_1, \ldots, A_n, E_i, F_k^i \hookrightarrow C \tag{13}$$

To capture exceptions to the exceptions we have to revise (11) to

$$A_1 \wedge \ldots \wedge A_n \wedge \bigwedge_{i=1}^{m} \left(\neg E_i \wedge \bigvee_{j=1}^{k} F_j^i \right) \rightarrow C \tag{14}$$

with the same problems of factual omniscience as above (and recursive refinement in case of exceptions to the exceptions to the exceptions). In contrast in defeasible reasoning, exceptions can simply be represented by replacing rules (9)–(10) with

$$E_1 \hookrightarrow \neg C \tag{15}$$

$$\ldots$$

$$E_m \hookrightarrow \neg C \tag{16}$$

plus a mechanism to solve conflict. The key point is that unless one is able to conclude the conditions of applicability of a rule, the rule does not fire, and it is not able to produce its conclusion.

In this section we gave a short overview of the salient features for practical legal reasoning, in the next section we are going to present how to implement them in a computationally oriented formalism: Defeasible Deontic Logic.

3   Defeasible Deontic Logic

Defeasible Deontic Logic is an extension of Defeasible Logic with deontic operators, and the operators for compensatory obligation introduced in [21]. Defeasible Logic is a simple rule based computationally oriented (and efficient) non-monotonic formalism, designed for handling exception in a natural way. While Defeasible Logic was originally proposed by Nute [29], we follow the formalisation proposed in [2]. Defeasible Logic is a constructive logic with its proof theory and inference condition as its core. The logic exploits both positive proofs, a conclusion has been constructively prove using the given rules and inference conditions (also called proof conditions), and negative proofs: showing a constructive and systematic failure of reaching particular conclusions, or in other terms, constructive refutations. The logic uses a simple language, that proved successful in many application area, due to the scalability of the logic, and its constructiveness. These elements are extremely important for normative reasoning, where an answer to a verdict is often nor enough, and full traceability is needed. This feature is provided by the constructive proof theory of Defeasible logic.

In the rest of this section we first provide an informal presentation of basic Defeasible Logic, we informally discuss how the extend it to cover the normative features we outlined in the previous section, and then finally, we provide the formal presentation of Defeasible Deontic Logic.

3.1   Basic Defeasible Logic

Knowledge in Defeasible logic is structured in three components:

- A set of facts (corresponding to indisputable statements represented as literals, where a literal is either an atomic proposition or its negation).
- A set of rules. A rule establishes a connection between a set of premises and a conclusion. In particular, for reasoning with norms, it is reasonable to assume that a rule provides the formal representation of a norm. Accordingly, the premises encode the conditions under which the norm is applicable, and the conclusion is the normative effect of the norm.
- A preference relation over the rules. The preference relation just gives the relative strength of rules. It is used in contexts where two rules with opposite conclusions fire simultaneously, and determines that one rule overrides the other in that particular context.

Formally, the knowledge in the logic is organised in Defeasible Theories, where a Defeasible Theory D is a structure

$$(F, R, \prec) \tag{17}$$

where F is the set of facts, R is the set of rules, and $\prec$ is a binary relation over the set of rules, i.e., $\prec \subseteq R \times R$.[6]

---

[6] Defeasible Logic does not impose any property for $\prec$. However, in many applications it is useful to assume that the transitive closure to be acyclic to prevent situations where, at the same time a rule overrules another rule and it is overridden by it.

As we have alluded to above, a rule is formally a binary relation between a set premises and a conclusion. Thus, if Lit is the set of literals, the set Rule of all rules is:

$$\text{Rule} \subseteq 2^{\text{Lit}} \times \text{Lit}. \tag{18}$$

Accordingly, a rule is an expression with the following form:[7]

$$r\colon a_1, \ldots, a_n \hookrightarrow c \tag{19}$$

where r is a unique label identifying the rule. Given that a rule is a relation, we can ask what is the strength of the link between the premises and the conclusion. We can distinguish three different strengths: (i) given the premises the conclusion always holds, (ii) given the premises the conclusion holds sometimes, and (iii) given the premises the opposite of the conclusions does not hold. Therefore, to capture theses types Defeasible Logic is equipped with three types of rules: *strict rules*, *defeasible rules* and *defeaters*. We will use $\rightarrow$, $\Rightarrow$ and $\rightsquigarrow$ instead of $\hookrightarrow$ to represent, respectively, strict rules, defeasible rules and defeaters.

Given a rule like rule r in (19) we use the following notation to refer to the various elements of the rule. $A(r)$ denotes the *antecedent* or *premises* of the rule, in this case, $\{a_1, \ldots, a_n\}$, and $C(r)$ denotes the *conclusion* or *consequent*, that is, c. From time to time we use *head* and *body* of a rule to refer, respectively, to the consequent and to the antecedent of the rule.

Strict rules are rules in the classic sense: whenever the premises are indisputable so is the conclusion. Strict rules can be used to model legal definitions that do not admit exceptions, for example the definition of minor: "'minor' means any person under the age of eighteen years". This definition can be represented as

$$age(x) < 18yrs \rightarrow minor(x). \tag{20}$$

Defeasible Rules are rules such that the conclusions normally or typically follow from the premises, unless there are evidence or reasons to the contrary.

Defeaters are rules that do not support directly the derivation of a conclusion, but that can be used to prevent a conclusion.

Finally, for the *superiority relation*, given two rules r and s, we use $r \prec s$ to indicate that rule r defeats rule s; in other terms, if the two rules are in conflict with each other and they are both applicable, then r prevails over s, and we derive only the conclusion of r.

Example 3. We illustrate defeasible rules and defeaters with the help of the definition of complaint from the Australian Telecommunication Consumer Protections Code 2012 TCP-C268_2012 May 2012 (TCPC).

> **Complaint** means an expression of dissatisfaction made to a Supplier in relation to its Telecommunications Products or the complaints handling process itself, where a response or Resolution is explicitly or implicitly expected by the Consumer.

---

[7] More correctly, we should use $r\colon \{a_1, \ldots, a_n\} \hookrightarrow c$. However, to improve readability, we drop the set notation for the antecedent of rule.

An initial call to a provider to request a service or information or to request support is not necessarily a Complaint. An initial call to report a fault or service difficulty is not a Complaint. However, if a Customer advises that they want this initial call treated as a Complaint, the Supplier will also treat this initial call as a Complaint.

If a Supplier is uncertain, a Supplier must ask a Customer if they wish to make a Complaint and must rely on the Customer's response.

Here is a (simplified) formal representation:

$$tcpc_1: ExpressionDissatisfaction \Rightarrow Complaint$$
$$tcpc_2: InformationCall \Rightarrow \neg Complaint$$
$$tcpc_3: ProblemCall, FirstCall \rightsquigarrow Complaint$$
$$tcpc_4: AdviseComplaint \Rightarrow Complaint$$

where $tcpc_2 \prec tcpc_1$ and $tcpc_4 \prec tcpc_2$.

The first rule $tcpc_1$ sets the basic conditions for something to be a complaint. On the other hand, rule $tcpc_2$ provides an exception to the first rule, and rule $tcpc_4$ is an exception to the exception provided by rule $tcpc_2$. Finally, $tcpc_3$ does not alone warrant the call to be a complaint (though, it does not preclude the possibility that the call turns out to be a complaint; hence the use of a defeater to capture this case).

Defeasible Logic is a constructive logic. This means that at the heart of it we have its proof theory, and for every conclusion we draw from a defeasible theory we can provide a proof for it, giving the steps used to reach the conclusion, and at the same time, providing a (formal) explanation or justification of the conclusion. Furthermore, the logic distinguishes *positive* and *negative* conclusion, and the strength of a conclusion. This is achieved by labelling each step in a derivation with a proof tag. As usual a derivation is a (finite) sequence of formulas, each obtained from the previous ones using inference conditions.

Let D be a Defeasible Theory. The following are the proof tags we consider for basic Defeasible Logic:

$+\Delta$ if a literal p is tagged by $+\Delta$, then this means that p is provable using only the facts and strict rules in a defeasible theory. We also say that p is *definitely provable* from D.

$-\Delta$ if a literal p is tagged by $-\Delta$, then this means that p is refuted using only the facts and strict rules in a defeasible theory. In other terms, it indicates that the literal p cannot be proved from D using only facts and strict rules. We also say that p is *definitely refuted* from D.

$+\partial$ if a literal p is tagged by $+\partial$, then this means that p is *defeasibly provable* from D.

$-\partial$ if a literal p is tagged by $-\partial$, then this means that p is *defeasibly refutable* from D.

Some more notation is needed before explaining how tagged conclusions can be asserted. Given a set of rules R, we use R_x to indicate particular subsets of rules: R_s for strict rules, R_d for defeasible rules, R_{sd} for strict or defeasible rules, R_{dft} for defeaters; finally $R[q]$ denotes the rules in R whose conclusion is q.

There are two ways to prove $+\Delta p$ at the n-th step of a derivation: the first is that p is one of the facts of the theory. The second case is when we have a strict rule r for p and all elements in the antecedent of r have been definitely proved at previous steps of the derivation.

For $-\Delta p$ we have to argue that there is no possible way to derive p using facts and strict rules. Accordingly, p must not be one of the facts of the theory, and second for every rule in $R_s[p]$ (all strict rules which are able to conclude p) the rule cannot be applied, meaning that at least one of the elements in the antecedent of the rule has already refuted (definitely refuted). The base case is where the literal to be refuted is not a fact and there are no strict rules having the literal as their head.

Defeasible derivations have a three phases argumentation-like structure[8]. To show that $+\partial p$ is provable at step n of a derivation we have to:[9]

1. give an argument for p;
2. consider all counterarguments for p; and
3. rebut each counterargument by either:
   (a) showing that the counterargument is not valid;
   (b) providing a valid argument for p defeating the counterargument.

In this context, in the first phase, an argument is simply a strict or defeasible rule for the conclusion we want to prove, where all the elements are at least defeasibly provable. In the second phase we consider all rules for the opposite or complement of the conclusion to be proved. Here, an argument (counterargument) is not valid if the argument is not supported.[10] Here "supported" means that all the elements of the body are at least defeasibly provable.

Finally to defeasibly refute a literal, we have to show that either, the opposite is at least defeasible provable, or show that an exhaustive search for a constructive proof for the literal fails (i.e., there are rules for such a conclusion or all rules

---

[8] The relationships between Defeasible Logic and argumentation are, in fact, deeper than the similarity of the argumentation like proof theory. [18] prove characterisation theorems for defeasible logic variants and Dung style argumentation semantics [7]. In addition, [11] proved that the Carneades argumentation framework [9], widely discussed in the AI and Law literature, turns out to be just a syntactic variant of Defeasible Logic.

[9] Here we concentrate on proper defeasible derivations. In addition we notice that a defeasible derivations inherit from definite derivations, thus we can assert $+\partial p$ if we have already established $+\Delta p$.

[10] It is possible to give different definitions of support to obtain variants of the logic tailored for various intuitions of non-monotonic reasoning. [4] show how to modify the notion of support to obtain variants capturing such intuitions, for example by weakening the requirements for a rule to be supported: instead of being defeasibly provable a rule is supported if it is possible to build a reasoning chain from the facts ignoring rules for the complements.

are either "invalid" argument or they are not stronger than valid arguments for the opposite).

Example 4. Consider again the set of rules encoding the TCPC 2012 definition of complaint given in Example 3. Assume to have a situation where there is an initial call from a customer who is dissatisfied with some aspects of the service received so far where she asks for some information about the service. In this case rules $tcpc_1$ and $tcpc_2$ are both applicable (we assume that the facts of the case include the union of the premises of the two rules, but *AdviseComplaint* is not a fact). Here, $tcpc_2$ defeats $tcpc_1$, and $tcpc_4$ cannot be used. Hence, we can conclude $-\partial AdviseComplaint$ and consequently $+\partial\neg Complaint$ and $-\partial Complaint$. However, if the customer stated that she wanted to complain for the service, then the fact *AdviseComplaint* would appears in the facts. Therefore, we can conclude $+\partial AdviseComplaint$, making then rule $tcpc_4$ applicable, and we can reverse the conclusions, namely: $+\partial Complaint$ and $-\partial\neg Complaint$.

While the Defeasible Logic we outlined in this section and its variants are able to model different features of legal reasoning (e.g., burden of proof [24] and proof standards [11] covering and extending the proof standards discussed in [9]), we believe that a few important characteristics of legal reasoning are missing. First, we do not address the temporal dimension of norms (and, obviously, this is of paramount importance to model norm dynamics), and second, we do not handle the normative character of norms: norms specify what are the obligations, prohibitions and permissions in force and what are the conditions under which they are in force. In the next sections we are going to extend Defeasible Logic with (1) deontic operators, to capture the normative nature of norms and (2) time, to model the temporal dimensions used in reasoning with norms.

3.2   The Intuitions Behind Defeasible Deontic Logic

In the language of Defeasible Deontic Logic the set of literals Lit is partitioned in *plain literals* and *deontic literals*. A plain literal is a literal in the sense of basic defeasible logic, while a deontic literal is obtained by placing a plain literal in the scope of a deontic operator or a negated deontic operator. Accordingly, expressions like Ol, ¬Pl and F¬l are deontic literals, where l is plain literal. We use ModLit to indicate the set of all deontic literals. We have now to give the construction for the operator for compensatory obligations ($\otimes$). We will refer to such expression as $\otimes$-expressions, whose construction rules are as follows:

(a) every literal $l \in$ Lit is an $\otimes$-expression;
(b) if $l_1, \ldots, l_n$ are plain literals, then $l_1 \otimes \cdots \otimes l_n$ is an $\otimes$-expression;
(c) nothing else is an $\otimes$-expression;
(d) the set of all $\otimes$-expression is denoted by Oexpr.

The main idea is that instead of a single family or rules, we are going to have to have two families: one for constitutive rules, and the other for prescriptive rules. The behaviour of these rules is different. Given the constitutive defeasible rule

$$a_1, \ldots, a_n \Rightarrow_C b \tag{21}$$

we can assert b, given $a_1, \ldots, a_n$, thus the behaviour of constitutive rule is just the normal behaviour of rules we examined in the previous section. Thus, when the rule is applicable, we can derive $+\partial_C b$ meaning that b is derivable. For prescriptive rules the behaviour is a different. From the rule

$$a_1, \ldots, a_n \Rightarrow_O b \tag{22}$$

we conclude $+\partial_O b$ meaning that we derive Ob when we have $a_1, \ldots, a_n$. Thus we conclude the obligation of the consequent of the rule, namely that b is obligatory, i.e., Ob, not just the consequent of the rule, i.e., b.[11]. Furthermore, for prescriptive rules we can have rules like

$$a_1, \ldots, a_n \Rightarrow_O b_1 \otimes \cdots \otimes b_m \tag{23}$$

In this case, when the rule is applicable, we derive, Ob_1 $(+\partial_O b_1)$, but if in addition we prove that $\neg b_1$ holds, we can conclude that we have a violation, and then we derive the compensatory obligation Ob_2, and so on.

The reasoning mechanism is essentially the same as that of basic defeasible presented in Sect. 3.1. The first difference is that an argument can only be attacked by an argument of the same type. Thus if we have an argument consisting of a constitutive rule for p, a counterargument should be a constitutive rule for $\sim p$. The same applies for prescriptive rule. An exception to this is when we have a constitutive rule for p such that all its premises are provable as obligations. In this case the constitutive rule behaves like a prescriptive rule, and can be used as a counterargument for a prescriptive rule for $\sim p$, or the other way around.

Consider, for example, the following two rules

$$r_1 : a_1, a_2 \Rightarrow_C b \tag{24}$$

$$r_2 : c \Rightarrow_O \neg b. \tag{25}$$

The idea expressed by r_1 is that, in a particular normative system, the combination of a_1 and a_2 is recognised as the institutional fact b, while r_2 prohibits b given c. Suppose now that a_1 and a_2 are both obligatory. Under these conditions it is admissible to assert that b is obligatory as well. Accordingly, r_1 can be used to conclude Ob instead of simply b. This means that the conclusions of r_1 and r_2 are conflicting: thus r_1, when its premises are asserted as obligation, can be used to counter an argument (e.g., r_2) forbidding b (making $\neg b$ obligatory, or $O \neg b$).

---

[11] As explained elsewhere [10,22], we do not add a deontic operator in the consequent of rules (i.e., $a_1, \ldots, a_n \Rightarrow Ob$), but we rather differentiate the mode of conclusions by distinguishing diverse rule types. This choice has a technical motivation: (a) it considerably makes simpler and more compact the proof theory; (b) it allows us to characterise a specific logical consequence relation for O, and eventually implement different proof conditions for constitutive rule, and prescriptive rules, also, to account for different burden of proof, and compliance and violations based on the definition of burden presented in [11,12,24] justified by the results by [17].

The second difference is that now the proof tags are labelled with either C, e.g., $+\partial_C p$, (for constitutive conclusions) or with O, e.g., $-\partial_O q$ (for prescriptive conclusions). Accordingly, when we are able to derive $+\partial_O p$ we can say that Op is (defeasibly) provable.

This feature poses the question of how we model the other deontic operators (i.e., permission and prohibition). As customary in Deontic Logic, we assume the following principles governing the interactions of the deontic operators.[12]

$$O{\sim}l \equiv \mathsf{F}l \tag{26}$$

$$Ol \land O{\sim}l \to \bot \tag{27}$$

$$Ol \land P{\sim}l \to \bot \tag{28}$$

Principle (26) provides the equivalence of a prohibition with a negative obligation (i.e., obligation not). The second and the third are rationality postulates stipulating that it is not possible to have that something and its opposite are at the same time obligatory (27) and that a normative system makes something obligatory and its opposite is permitted (28). (26) gives us the immediate answer on how prohibition is modelled. A rule giving a prohibition can be modelled just as a prescriptive rule for a negated literal. This means that to conclude $\mathsf{F}p$ we have to derive $+\partial_O \neg p$, in other terms that $\neg p$ is (defeasibly) provable as an obligation.

Example 5. Section. 40 of the Australian Road Rules (ARR)[13].

Making a U–turn at an intersection with traffic lights
A driver must not make a U–turn at an intersection with traffic lights unless there is a U–turn permitted sign at the intersection.

The prohibition of making U-turns at traffic lights can be encoded by the following rule:

$$arr_{40a}\colon AtTrafficLigths \Rightarrow_O \neg Uturn.$$

In a situation where *AtTrafficLights* is given we derive $+\partial_O \neg Uturn$ which corresponds to F *Uturn*.

The pending issue is how to model permissions. Two types of permissions have been discussed in literature following [1,34,35]: (i) weak permission, meaning

---

[12] In the three formulas below $\to$ is the material implication of classical logic.

[13] This norm makes use of "must not", to see that "must not" is understood as prohibition in legal documents see, the Australian National Consumer Credit Protection Act 2009, Sect. 29, whose heading is "Prohibition on engaging in credit activities without a licence", recites "(1) A person must not engage in a credit activity if the person does not hold a licence authorising the person to engage in the credit activity".

that there is no obligation to the contrary; and (ii) strong permission, a permission explicitly derogates an obligation to the contrary. In this case we have an exception. For both types of permission we have that the obligation to the contrary does not hold. Defeasible Deontic Logic is capable to handle the two types of permission is a single shot if we establish that Pp is captured by $-\partial_O \sim p$. The meaning of $-\partial_O p$ is that p is refuted as obligation, or that it is not possible to prove p as an obligation; hence it means that we cannot establish that p is obligatory, thus there is no obligation contrary to $\sim p$.

The final aspect we address is how to model strong permissions. Remember that strong permissions are meant to be exceptions. Exceptions in Defeasible Logic can be easily captured by rules for the opposite plus a superiority relation. Accordingly, this could be modelled by

$$arr_{40e}: UturnPermittedSign \Rightarrow_O Uturn.$$

and $arr_{40e} \prec arr_{40a}$. We use a prescriptive defeasible rule for obligation to block the prohibition to U-turn. But, since arr_{49e} prevails over arr_{49a}, we derive that U-turn is obligatory, i.e., $+\partial_O Uturn$.

Thus, when permissions derogate to prohibitions (or obligations), there are good reasons to argue that defeaters for O are suitable to express an idea of strong permission[14]. Explicit rules such as $r: a \rightsquigarrow_O q$ state that a is a specific reason for blocking the derivation of $O\neg q$ (but not for proving Oq), i.e., this rule does not support any conclusion, but states that $\neg q$ is deontically undesirable. Accordingly, we can rewrite the derogating rule as

$$arr_{40e}: UturnPermittedSign \rightsquigarrow_O Uturn.$$

In this case, given $UturnPermittedSign$ we derive $-\partial_O \neg Uturn$.

3.3   Defeasible Deontic Logic Formalised

The presentation in this section is based on [19], and while it is possible to use the standard definitions of *strict rules*, *defeasible rules*, and *defeaters* [2] and the full language presented in [19] for the sake of simplicity, and to better focus on the non-monotonic and deontic aspects, we restrict our attention to defeasible rules and defeaters, and we assume that facts are restricted to plain literals. In addition, we use the equivalences between obligations and prohibitions to transform all prohibitions in the corresponding obligations.

A Defeasible Deontic Theory is a structure

$$(F, R^C, R^O, \prec) \tag{29}$$

where F is a set of facts, where every R^C is a set of constitutive rules, where a constitutive rule obeys the following signature

$$2^{Lit} \times PLit \tag{30}$$

---

[14] The idea of using defeaters to introduce permissions was introduced by [23].

while the signature for R^O, a set of prescriptive rules is

$$2^{\text{Lit}} \times \text{Oexpr} \tag{31}$$

Notice that in both cases, the antecedent of a rule can contain both plain and deontic literal, but, in any case, the conclusion is plain literal. Thus, the question is if the conclusions of rules are plain literals, where do we get deontic literals? The answer is that we have two different modes of for the rules. The first mode is that of constitutive rule, where the conclusion is an assertion with the same mode as it appears in the rule (i.e., as an institutional fact); the second mode is that of prescriptive rule, where the conclusion is asserted with a deontic mode (where the deontic mode corresponds to one of the deontic operators).

Constitutive rules behaves as the rules in Basic Defeasible Logic, and we use $\hookrightarrow_C$ to denote the arrow of a constitutive rule. $\hookrightarrow_O$ for the arrow of a prescriptive rule.

For $\otimes$-expression we stipulate that they obey the following properties:

1. $a \otimes (b \otimes c) = (a \otimes b) \otimes c$ (associativity);
2. $\bigotimes_{i=1}^{n} a_i = (\bigotimes_{i=1}^{k-1} a_i) \otimes (\bigotimes_{i=k+1}^{n} a_i)$ where there exists j such that $a_j = a_k$ and $j < k$ (duplication and contraction on the right).

Given an $\otimes$-expression A, the *length* of A is the number of literals in it. Given an $\otimes$-expression $A \otimes b \otimes C$ (where A and C can be empty), the *index* of b is the length of $A \otimes b$. We also say that b appears at index n in $A \otimes b$ if the length of $A \otimes b$ is n.

Given a set of rules R, we use the following abbreviations for specific subsets of rules:

- R_{def} denotes the set of all defeaters in the set R;
- $R[q, n]$ is the set of rules where q appears at index n in the consequent. The set of (defeasible) rules where q appears at any index n is denoted by $R[q]$;
- $R^O[q, n]$ is the set of (defeasible) rules where q appears at index n and the operator preceding it is $\otimes$ for $n > 1$ or the mode of the rule is O for $n = 1$. The set of (defeasible) rules where q appears at any index n is denoted by $R^O[q]$;

A *proof* P in a defeasible theory D is a linear sequence $P(1) \dots P(n)$ of *tagged literals* in the form of $+\partial_\square q$ and $-\partial_\square q$ with $\square \in \{C, O, P, P_w, P_s\}$, where $P(1) \dots P(n)$ satisfy the proof conditions given in the rest of the paper. The initial part of length i of a proof P is denoted by $P(1...i)$. The tagged literal $+\partial_\square q$ means that q is *defeasibly provable* in D with modality $\square$, while $-\partial_\square q$ means that q is *defeasibly refuted* with modality $\square$. More specifically we can establish the following relationships

- $+\partial_C a$: a is provable;
- $-\partial_C a$: a is rejected;
- $+\partial_O a$: Oa is provable;
- $+\partial_O \neg a$: Fa is provable;

- $-\partial_O a$: Oa is rejected;
- $-\partial_O \neg a$: Fa is rejected;
- $+\partial_P a$: Pa is provable;
- $-\partial_P a$: Pa is rejected;
- $+\partial_{P_s} a$: $P_s a$ is provable;
- $-\partial_{P_s} a$: $P_s a$ is rejected;
- $+\partial_{P_w} a$: $P_w a$ is provable;
- $-\partial_{P_w} a$: $P_w a$ is rejected;

The first thing to do is to define when a rule is applicable or discarded. A rule is *applicable* for a literal q if q occurs in the head of the rule, all non-modal literals in the antecedent are given as facts and all the modal literals have been defeasibly proved (with the appropriate modalities). On the other hand, a rule is *discarded* if at least one of the modal literals in the antecedent has not been proved (or is not a fact in the case of non-modal literals). However, as literal q might not appear as the first element in an $\otimes$-expression in the head of the rule, some additional conditions on the consequent of rules must be satisfied. Defining when a rule is applicable or discarded is essential to characterise the notion of provability for obligations ($\pm\partial_O$) and permissions ($\pm\partial_P$).

A rule $r \in R[q, j]$ is *body-applicable* iff for all $a_i \in A(r)$:

1. if $a_i = \Box l$ then $+\partial_\Box l \in P(1...n)$ with $\Box \in \{C, O, P, P_w, P_s\}$;
2. if $a_i = \neg\Box l$ then $-\partial_\Box l \in P(1...n)$ with $\Box \in \{C, O, P, P_w, P_s\}$;
3. if $a_i = l \in \mathrm{Lit}$ then $l \in F$.

A rule $r \in R[q, j]$ is *body-discarded* iff $\exists a_i \in A(r)$ such that

1. if $a_i = \Box l$ then $-\partial_\Box l \in P(1...n)$ with $\Box \in \{C, O, P, P_w, P_s\}$;
2. if $a_i = \neg\Box l$ then $+\partial_\Box l \in P(1...n)$ with $\Box \in \{C, O, P, P_w, P_s\}$;
3. if $a_i = l \in \mathrm{Lit}$ then $l \notin F$.

A rule $r \in R$ is *body-p-applicable* iff

1. if $r \in R^O$ and it is body-applicable; or
2. if $r \in R^C$ and, $A(r) \neq \emptyset$, $A(r) \subseteq \mathrm{PLit}$ and $\forall a_i \in A(r)$, $+\partial_O a_i \in P(1...n)$.

A rule $r \in R$ is *body-p-discarded* iff

1. if $r \in R^O$ and it is not body-applicable; or
2. if $r \in R^C$ and either $A(r) = \emptyset$ or $A(r) \cap \mathrm{DLit} \neq \emptyset$ and $\exists a_i \in A(r)$, $-\partial_O a_i \in P(1...n)$.

The last two conditions are used to determine if we can use a constitutive rule to derive an obligation when all the elements in the antecedent are provable as an obligation but in the constitutive rule require them to be factually derivable, see the discussion for (24) above.

A rule $r \in R[q, j]$ such that $C(r) = c_1 \otimes \cdots \otimes c_n$ is *applicable* for literal q at index j, with $1 \leq j < n$, in the condition for $\pm\partial_O$ iff

1. r is body-p-applicable; and
2. for all $c_k \in C(r)$, $1 \leq k < j$, $+\partial_O c_k \in P(1...n)$ and $(c_k \notin F$ or $\sim c_k \in F)$.

Conditions (1) represents the requirements on the antecedent while condition (2) on the head of the rule states that each element c_k prior to q must be derived as an obligation, and a violation of such obligation has occurred.

We are now ready to give the proof conditions for the proof tags used in Defeasible Deontic Logic. The conditions for $+\partial_\Box$ and $-\partial_\Box$ are related by the Principle of Strong Negation [3,20]: The *strong negation* of a formula is closely related to the function that simplifies a formula by moving all negations to an inner most position in the resulting formula, and replaces the positive tags with respective negative tags, and vice versa. We will give the positive and negative proof conditions for one tag to illustrate how the principle works. For the remaining tag we present only the positive case.

The proof condition of *defeasible provability for a factual conclusion* is

$+\partial_C$: If $P(n + 1) = +\partial_C q$ then
(1) $q \in F$ or
    (2.1) $\sim q \notin F$ and
    (2.2) $\exists r \in R^C[q]$ such that r is body-applicable, and
    (2.3) $\forall s \in R[\sim q]$, either
        (2.3.1) s is body-discarded, or
        (2.3.2) $\exists t \in R^O[q]$ such that t is body-applicable and $s \prec t$.

The proof condition for *defeasible refutability for a factual conclusion* is

$-\partial_C$: If $P(n + 1) = -\partial_C q$ then
(1) $q \notin F$ and
    (2.1) $\sim q \in F$ or
    (2.2) $\forall r \in R^C[q]$ either r is body-discarded, or
    (2.3) $\exists s \in R[\sim q]$, such that
        (2.3.1) s is body-applicable for q, and
        (2.3.2) $\forall t \in R^O[q]$ either t is body-discarded or not $s \prec t$.

The proof conditions for $\pm\partial_C$ are the standard conditions of basic defeasible logic with the proviso that the element of the antecedents of the rules are applicable in the extended deontic logic.

The proof condition of *defeasible provability for obligation* is

$+\partial_O$: If $P(n + 1) = +\partial_O q$ then
(1) $\exists r \in R_d^O[q, i] \cup R_d^C[q]$ such that r is applicable for q, and
(2) $\forall s \in R[\sim q, j]$, either
    (2.1) s is discarded, or
    (2.2) $\exists t \in R[q, k]$ such that t is applicable for q and $s \prec t$.

To show that q is defeasibly provable as an obligation, then q must be derived the prescriptive rules or by rules of the theory that behaves as prescriptive rules in the theory. More specifically, (1) there must be a rule introducing the obligation for q which can apply; this is possible if the rule is a prescriptive rules and all the applicability conditions hold, or there is a constitutive rule where all the conditions of applicability are mandatory (i.e., are provable as obligations).

(2) every rule s for $\sim q$ is either discarded or defeated by a stronger rule for q. If s is an obligation rule, then it can be counterattacked by any type of rule.

The strong negation of the condition above gives the negative proof condition for obligation.

Let us now move to the proof conditions for permission. We start with the condition for strong permission.

The proof condition of *defeasible provability for strong permission* is

$+\partial_{P_s}$: If $P(n+1) = +\partial_{P_s}q$ then
(1) $+\partial_O q \in P(1...n)$ or
 (2.1) $\exists r \in R^O_{def}[q] \cup R^C_{def}[q]$ such that r is body-p-applicable, and
 (2.2) $\forall s \in R[\sim q, j]$, either
  (2.2.1) s is discarded, or
  (2.2.2) $\exists t \in R[q, k]$ such that t is applicable for q and $s \prec t$.

In this case we have two conditions. The first condition is that, we inherit the strong permission from an obligation (see the discussion in Sect. 2). In the second case, the condition has the same structure as that of a defeasible obligation, but instead of a defeasible rule there is a body-applicable prescriptive defeater or body-p-applicable constitutive defeater, that is not defeated (see the discussion in Sect. 3.2, in particular the formalisation of the U-turn at traffic lights example).

The proof condition for weak permission simply boils down to the failure to prove the obligation to the contrary, namely:

$+\partial_{P_w}$: If $P(n+1) = +\partial_{P_w}q$ then
(1) $-\partial_O \sim q \in P(1...n)$.

Similarly, for a generic permission, a permission that does not distinguish between strong and weak permission, the proof condition is that we have proved either to the two:

$+\partial_P$: If $P(n+1) = +\partial_P q$ then
(1) $+\partial_{P_s}q \in P(1...n)$ or
(2) $+\partial_{P_w}q \in P(1...n)$.

Notice that in the majority of cases, where a permission appears in the antecedent of a rule the permission is understood as a generic permission.

Finally, we introduce a special proof tag for a distinguished literal meant to represent an non-compensable violation. To this end we extend the set of literal language with the proposition $\bot$. Furthermore, we stipulate that this literal can only appear in the antecedent of rules. As we have seen we can use $\otimes$-expressions to model obligations, their violations, and the corresponding compensatory obligations. Thus, given the $\otimes$-expression

$$a_1 \otimes \cdots \otimes a_n \tag{32}$$

Oa_1 is the ideal primary obligations, Oa_2 is the second best option when there is a violation of the primary obligation, and if we repeat this argument, a_n is the least one can do to compensate the violations that precede it, and no further compensation is then possible. Thus, we can say that we have obligation that cannot be compensated if we fully traverse an $\otimes$-expression. Thus, we can say that we have a violation that cannot be compensate if the following proof condition holds.

$+\partial_\perp$: If $P(n+1) = +\partial_\perp$ then

(1) $\exists R^O{}_d \cup R^C_d$ such that

    (1.1) r is body-p-applicable and

    (1.2) $\forall c_i \in C(r) +\partial_O c_i \in P(1...n)$ and either $c_i \notin F$ or $\sim c_i \in F$.

We conclude this section by reporting some results about the logic, showing the behaviour of the logic, its deontic and computational properties. In what follows, given a defeasible deontic theory D, we use $D \vdash \pm\#q$, where $\#$ stand for any of the proof tags discussed in this section, to denote that there is a derivation of $\#q$ from D.

Proposition 1. *Given a defeasible deontic theory D.*

1. *it is not possible to have both $D \vdash +\#q$ and $D \vdash -\#q$;*
2. *$D \vdash +\partial_C q$ and $D \vdash +\partial_C \neg q$ if $q, \neg q \in F$;*
3. *for $\Box \in \{O, P_s\}$, it is not possible to have both $D \vdash +\partial_\Box q$ and $D \vdash +\partial_\Box \sim q$;*
4. *for $\Box \in \{O, P_s\}$, if $D \vdash +\partial_\Box q$, then $D \vdash -\partial_\Box \sim q$;*
5. *it is not possible to have both $D \vdash +\partial_O q$ and $D \vdash +\partial_{P_s} \sim q$;*
6. *for $\Box \in \{P, P_s, P_w\}$, if $D \vdash +\partial_O q$, then $D \vdash +\Box q$;*
7. *if $D \vdash +\partial_{P_w} q$, then $D \vdash -\partial_O \sim q$.*

The first property is an immediate consequence of using the Principle of Strong Negation in formulating the proof conditions. Properties 2 and 3 establish the consistency of the logic. The inference mechanisms cannot produce an inconsistency unless the monotonic part (facts) is already inconsistent. Property 4 is again a corollary of the Principle of Strong Negation. Finally, Properties 5 and 6 show that the deontic operators satisfy the properties normally ascribed to them in deontic logics.

Given a Defeasible Theory D, HB_D is the set of literals such that the literal or its complement appears in D, where 'appears' means that it is a sub-formula of a modal literal occurring in the theory. The modal Herbrand Base of D is $HB = \{\Box l | \Box \in \{C, O, P, P_w, P_s\}, l \in HB_D\}$. Accordingly, the extension of a Defeasible Theory is defined as follows.

Given a Defeasible Theory D, the *defeasible extension* of D is defined as

$$E(D) = (+\partial_\Box, -\partial_\Box)$$

where $\pm\partial_\Box = \{l \in HB_D : D \vdash \pm\partial_\Box l\}$ with $\Box \in \{C, O, P, P_w, P_s\}$. In other words, the extension is the (finite) set of conclusions, in the Herbrand Base, that can be derived from the theory.

Proposition 2. *Given a defeasible theory D, the extension of the theory $E(D)$ can be computed in $O(size(D))$. Where the size of the theory D, $size(D)$ is determined by the number of occurrences of literals in the theory.*

The procedure to compute the extension in polynomial (linear) time is given in [19] and it is based on the algorithm given by Maher [28] to show that computing the extension of a basic defeasible theory can be computed in linear time.

4 Modelling Norms with Defeasible Deontic Logic

We begin this paper with a scenario, Example 1, that has been used to identify limitations of other approach to model real life norms (to be precise, it is an extension of the example to show the drawback of other approach to model norms using formalism based on possible world semantics, see [14, 15]). Let us see how to model it in Defeasible Deontic Logic.

The first step is to understood the intent of the contract. We can paraphrase the contract as follows:

C0: the use of the product is forbidden;

C1: if a license is granted, then the use of the product is permitted;

C2: the publication of the result of the evaluation is forbidden;

C2c: the removal of the illegally published results within the allotted times compensates the illegal publication;

C2e: the publication of the results of the evaluation is permitted if approval is obtained before publication;

C3: commenting about the evaluation is forbidden;

C3e: commenting about the evaluation is permitted, it publication of the results is permitted;

C4: publication of the results of evaluation is obligatory, if commissioned for an independent evaluation;

C4x: the use of product is obligatory, if commissioned for an independent evaluation;

C5: the use of the product is forbidden, if there is a violation of the above conditions.

Based on the decomposition above we can formulate the following rules and instances of the superiority relation:

$$r_0: \Rightarrow_O \neg use$$

$$r_1: license \leadsto_O use$$

$$r_2: \Rightarrow_O \neg publish \otimes remove$$

$$r_{2e}: approval \leadsto_O publish$$

$$r_3: \Rightarrow_O \neg comment$$

$$r_{3e}: \mathsf{P} publish \leadsto_O comment$$

$$r_4: commission \Rightarrow_O publish$$

$$r_{4x}: commission \Rightarrow_O use$$

$$r_5: \bot \Rightarrow_O \neg use$$

where $r_O \prec r_1$, $r_o \prec r_{4x}$, $r_1 \prec r_5$, $r_{4x} \prec r_5$ $r_2 \prec r_{2e}$, $r_3 \prec r_{3e}$.

The situation problematic for other approaches is the case, that the licensee, with a valid license, illegally publishes the results of the evaluation, remove them, and at the same time post a comment about the evaluation. The other approaches conclude that commenting is permitted, but the permission to post a comment requires the permission to publish, but since the publishing was illegal, there was not permission to publish. For us, if we assume, the facts *license*, *publish*, *remove* and *comment*, then from r_1 we derive P*use*, from r_2, we have F*publish*, and O*remove*. Given that we have F*publish* we cannot conclude P*publish*, thus, $-\partial_\text{P} publish$ hence from r_3, we obtain F*comment*, and thus the conditions to derive $+\partial_\bot$ indicating that the situation is not compliant.

5   Conclusion and Summary

We began this paper with some simple and small examples containing essential features for normative reasoning. We discussed, why defeasible reasoning, supplemented with deontic operators has the capability to offer a conceptually sound and computationally feasible tool for legal reasoning (or aspects of it). Then, we proposed a specific logic, Defeasible Deontic Logic, and we show how to capture the various features in the logic, we illustrated some of the feature with legal examples. The logic presented in the paper has a full implementation [27], and it has been used in a number of applications and industry scale pilot projects (for example to determine the regulatory compliance of business processes [13]). Recently, the logic has been proposed for the reasoning engine for a platform to handle legislations in a digital format in the Regulation as a Platform (RaaP) project funded by the Australian Federal Government. As part of the project, several Acts have been encoded in Defeasible Deontic Logic demonstrating the suitability of the logic to model normative systems. The Regulation as a Platform framework is available for public evaluation at https://raap.d61.io.

References

1. Alchourrón, C.E., Bulygin, E.: Permission and permissive norms. In: Krawietz, W., et al. (ed.) Theorie der Normen. Duncker & Humblot (1984)
2. Antoniou, G., Billington, D., Governatori, G., Maher, M.J.: Representation results for defeasible logic. ACM Trans. Comput. Log. **2**(2), 255–287 (2001)
3. Antoniou, G., Billington, D., Governatori, G., Maher, M.J., Rock, A.: A family of defeasible reasoning logics and its implementation. In: Horn, W. (ed.) In: Proceedings of the 14th European Conference on Artificial Intelligence, ECAI 2000, Amsterdam, pp. 459–463. IOS Press (2000)
4. Billington, D., Antoniou, G., Governatori, G., Maher, M.J.: An inclusion theorem for defeasible logic. ACM Trans. Comput. Log. **12**(1), 6 (2010)
5. Carmo, J., Jones, A.J.I.: Deontic logic and contrary-to-duties. In: Gabbay, D.M., Guenthner, F. (eds.) Handbook of Philosophical Logic, vol. 8, pp. 265–343. Springer, Dordrecht (2002). https://doi.org/10.1007/978-94-010-0387-2_4
6. Chisholm, R.M.: Contrary-to-duty imperatives and deontic logic. Analysis **24**, 33–36 (1963)
7. Dung, P.M.: On the acceptability of arguments and its fundamental role in non-monotonic reasoning, logic programming and n-person games. Artif. Intell. **77**(2), 321–358 (1995)

8. Gordon, T.F., Governatori, G., Rotolo, A.: Rules and norms: requirements for rule interchange languages in the legal domain. In: Governatori, G., Hall, J., Paschke, A. (eds.) RuleML 2009. LNCS, vol. 5858, pp. 282–296. Springer, Heidelberg (2009). https://doi.org/10.1007/978-3-642-04985-9_26

9. Gordon, T.F., Prakken, H., Walton, D.: The Carneades model of argument and burden of proof. Artif. Intell. **171**(10–11), 875–896 (2007)

10. Governatori, G.: Representing business contracts in RuleML. Int. J. Coop. Inf. Syst. **14**(2–3), 181–216 (2005)

11. Governatori, G.: On the relationship between Carneades and defeasible logic, pp. 31–40. ACM (2011)

12. Governatori, G.: Burden of compliance and burden of violations. In: Rotolo, A. (ed.) 28th Annual Conference on Legal Knowledge and Information Systems, Frontieres in Artificial Intelligence and Applications, Amsterdam, pp. 31–40. IOS Press (2015)

13. Governatori, G.: The Regorous approach to process compliance. In: 2015 IEEE 19th International Enterprise Distributed Object Computing Workshop, pp. 33–40. IEEE Press (2015)

14. Governatori, G.: Thou shalt is not you will. In: Atkinson, K. (ed.) Proceedings of the Fifteenth International Conference on Artificial Intelligence and Law, pp. 63–68. ACM, New York (2015)

15. Governatori, G., Hashmi, M.: No time for compliance. In: Hallé, S., Mayer, W. (ed.) 2015 IEEE 19th Enterprise Distributed Object Computing Conference, pp. 9–18. IEEE (2015)

16. Governatori, G., Idelberg, F., Milosevic, Z., Riveret, R., Sartor, G., Xu, X.: On legal contracts, imperative and declarative smartcontracts, and blockchain systems. Artif. Intell. Law 1–33 (2018)

17. Governatori, G., Maher, M.J.: Annotated defeasible logic. Theory Pract. Log. Program. **17**(5–6), 819–836 (2017)

18. Governatori, G., Maher, M.J., Billington, D., Antoniou, G.: Argumentation semantics for defeasible logics. J. Log. Comput. **14**(5), 675–702 (2004)

19. Governatori, G., Olivieri, F., Rotolo, A., Scannapieco, S.: Computing strong and weak permissions in defeasible logic. J. Philos. Log. **42**(6), 799–829 (2013)

20. Governatori, G., Padmanabhan, V., Rotolo, A., Sattar, A.: A defeasible logic for modelling policy-based intentions and motivational attitudes. Log. J. IGPL **17**(3), 227–265 (2009)

21. Governatori, G., Rotolo, A.: Logic of violations: a Gentzen system for reasoning with contrary-to-duty obligations. Australas. J. Log. **4**, 193–215 (2006)

22. Governatori, G., Rotolo, A.: Changing legal systems: legal abrogations and annulments in defeasible logic. Log. J. IGPL **18**(1), 157–194 (2010)

23. Governatori, G., Rotolo, A., Sartor, G.: Temporalised normative positions in defeasible logic. In: Gardner, A. (ed.) 10th International Conference on Artificial Intelligence and Law, pp. 25–34. ACM Press (2005)

24. Governatori, G., Sartor, G.: Burdens of proof in monological argumentation. In: Winkels, R. (ed.) The Twenty-Third Annual Conference on Legal Knowledge and Information Systems, Volume 223 of Frontiers in Artificial Intelligence and Applications, Amsterdam, pp. 57–66. IOS Press (2010)

25. Grossi, D., Jones, A.: Constitutive norms and counts-as conditionals. Handb. Deontic Log. Norm. Syst. **1**, 407–441 (2013)

26. Idelberger, F., Governatori, G., Riveret, R., Sartor, G.: Evaluation of logic-based smart contracts for blockchain systems. In: Alferes, J.J., Bertossi, L., Governatori, G., Fodor, P., Roman, D. (eds.) RuleML 2016. LNCS, vol. 9718, pp. 167–183. Springer, Cham (2016). https://doi.org/10.1007/978-3-319-42019-6_11

27. Lam, H.-P., Governatori, G.: The making of SPINdle. In: Governatori, G., Hall, J., Paschke, A. (eds.) RuleML 2009. LNCS, vol. 5858, pp. 315–322. Springer, Heidelberg (2009). https://doi.org/10.1007/978-3-642-04985-9_29

28. Maher, M.J.: Propositional defeasible logic has linear complexity. Theory Pract. Log. Program. **1**(6), 691–711 (2001)

29. Nute, D.: Defeasible logic. In: Gabbay, D.M., Hogger, C.J., Robinson, J.A. (eds.) Handbook of Logic in Artificial Intelligence and Logic Programming, vol. 3, pp. 353–395. Oxford University Press, Oxford (1994)

30. Palmirani, M., Governatori, G., Athan, T., Boley, H., Paschke, A., Wyner, A. (eds.) LegalRuleML Core Specification Version 1.0. OASIS (2017)

31. Prakken, H., Sergot, M.: Contrary-to-duty obligations. Studia Logica **57**(1), 91–115 (1996)

32. Sartor, G.: Legal Reasoning: A Cognitive Approach to the Law. Springer (2005)

33. Searle, J.R.: The Construction of Social Reality. The Free Press, New York (1996)

34. Soeteman, A.: Logic in Law. Kluwer, Dordrecht (1989)

35. von Wright, G.H.: Norm and Action: A Logical Inquiry. Routledge and Kegan Paul, Abingdon (1963)

Efficient and Convenient SPARQL+Text Search: A Quick Survey

Hannah Bast[✉] and Niklas Schnelle

University of Freiburg, 79110 Freiburg, Germany
{bast,schnelle}@cs.uni-freiburg.de

Abstract. This is a quick survey about efficient search on a text corpus combined with a knowledge base. We provide a high-level description of two systems for searching such data efficiently. The first and older system, Broccoli, provides a very convenient UI that can be used without expert knowledge of the underlying data. The price is a limited query language. The second and newer system, QLever, provides an efficient query engine for SPARQL+Text, an extension of SPARQL to text search. As an outlook, we discuss the question of how to provide a system with the power of QLever and the convenience of Broccoli. Both Broccoli and QLever are also useful when only searching a knowledge base (without additional text).

Keywords: Knowledge bases · Semantic search · SPARQL+Text
Efficiency · User interfaces

1 Introduction

This short survey is about efficient search on a text corpus combined with a knowledge base. For the purpose of this paper, a knowledge base is a collection of subject-predicate-object triples, like in the following example. Note that objects can also be strings, called *literals*, and that such literals can contain a qualifier indicating the language.

<Neil_Armstrong> <Space_Agency> <NASA>
<Neil_Armstrong> <Place_of_birth> <Wapakoneta>
<Neil_Armstrong> <Date_of_birth> "1930-08-05"
<NASA> <Slogan> "For the Benefit of All"@en

Knowledge bases are well suited for structured data, which has a natural representation in the form above. With information cast in this form, we can ask queries with precise semantics, just like on a database. Here is an example query, formulated in SPARQL, the standard query language on knowledge bases. The query asks for astronauts and their agencies. The result is a table with two columns. If an astronaut works for k agencies, the table has k rows for that astronaut.

© Springer Nature Switzerland AG 2018
C. d'Amato and M. Theobald (Eds.): Reasoning Web 2018, LNCS 11078, pp. 26–34, 2018.
https://doi.org/10.1007/978-3-030-00338-8_2

```
SELECT ?x ?y WHERE {
  ?x <is-a> <Astronaut>.
  ?x <Space_Agency> ?y
}
```

Note that it is crucial that in the knowledge base the same identifier is used to denote the same entity in different triples. This is easier said than done: in practice, knowledge bases are often co-productions of large teams of people, and it is not trivial to ensure that different people use the same identifier when referring to the same entity.

Much of today's information is available only in text form. The main reasons for this are as follows. First, text as a direct representation of spoken language is a very natural form of communication for humans and it requires extra effort to convert a given piece of information into a structured form. Second, as a particular consequence of the first reason, especially current and expert information is much more likely to be available in text form than in structured form. Third, not all information can be meaningfully cast in the above mentioned triple form. For example, consider the following sentence, which, among other pieces of information, expresses that Neil Armstrong walked on the moon:

On July 21st 1969, Neil Armstrong became the first man to walk on the Moon.

The statement that a certain person walked on the moon is rather specific and applies to only few entities. Casting this into triple form might be reasonable for statements of historical importance. Doing it for all statements that are mentioned in a large text corpus, would lead to a knowledge base that has hardly more structure than the text corpus itself. Also note that the larger the set of predicates becomes, the harder it becomes to maintain the above-mentioned property of using consistent identifiers.

It is therefore very natural to consider both knowledge bases and text for search. The simplest way to achieve this is to query both data sets separately and return the results as distinct result sets (provided that matches were found). This is essentially what the big commercial search engines currently do. In this paper, we consider a more powerful search, which considers the knowledge base and the text in combination. In the following, Sect. 2 explains the nature of this combination. Section 3 presents a system, called *Broccoli*, which enables a user to interactively and conveniently search on such a combined data set. Section 4 presents *SPARQL+Text*, a powerful extension of *SPARQL*, along with *QLever*, an efficient query engine implementing this extension. Section 5 briefly discusses how to combine the power of a system like QLever with the convenience of a system like Broccoli.

Both Broccoli and QLever are available online. A demo of Broccoli is available under http://broccoli.cs.uni-freiburg.de. A demo of QLever is available under http://qlever.cs.uni-freiburg.de. The source code of QLever is available under http://github.com/ad-freiburg/qlever.

2 Combining Text and Knowledge Base Data

A natural way to connect a given text corpus to a given knowledge base is to identify which pieces of text refer to an entity from the knowledge base, and exactly which entity is meant in each case. These problems are known as *named entity recognition* (NER) and *named entity disambiguation* (NED). For example, consider the following sentence:

Buzz Aldrin joined Armstrong and became the second human to set foot on the Moon.

Underlining the correct tokens is the NER problem. This entails figuring out the extent of the text used to refer to one entity. This can be just a single word, but it can also be a sequence of two or more words. Identifying which entities from the knowledge base the underlined pieces of text refer to is the NED problem. Note that there are many entities which could be referred to by the word *Armstrong*. Figuring out the correct entity can be a hard problem. In the sentence above, the other words in the sentence make it pretty clear that Neil Armstrong is meant. In the ERD'14 challenge [10], the best approach achieved an F-measure of 76% [12]. A quick overview of the state of the art in NER and NED can be found in a recent survey [7, Sect. 3.2.2]. In the following, we simply assume that the NER+NED problem has been solved satisfactorily and we thus have a text and a knowledge base linked in the way described.

With a text corpus and a knowledge base linked in this manner, we now have a notion of *co-occurrence* of an entity from the knowledge base with one or more words from the text corpus. The two systems described in Sects. 3 and 4 allow to specify such co-occurrences as part of a query. For example, both systems allow a query that searches for entities in the knowledge base with the profession *astronaut* (that is, astronauts), which somewhere in the text co-occur with a word starting with *walk* (walk, walked, walking, ...) and the word *moon*. The scope of the co-occurrence is an additional parameter: for example, we can consider co-occurrence within the same sentence or co-occurrence within the same grammatical sub-clause of a sentence; this is explained further in Sect. 3. Note how such a co-occurrence is a good indicator that the respective astronaut indeed walked on the moon. The more such co-occurrences we find, the more likely it is. The examples given in the next sections will clarify this further.

To get a feeling for the amount of data which these systems can handle, here are two concrete datasets, which have been used in the evaluation of these systems.

Wikipedia+Freebase Easy: The text from a dump of the English Wikipedia with links to a curated and simplified version of Freebase called *Freebase Easy* [2]. In particular, all entities and predicates in Freebase Easy are denoted by unique human-readable names, like in the example triples at the beginning of the introduction. The dataset is available at http://freebase-easy.cs.uni-freiburg. de. The dimensions of this dataset are: 360,744,363 triples, 2,316,712,760 word occurrences, 494,253,129 entity links.

Clueweb+Freebase: The text from the Clueweb12 collection [11] with links to the latest complete version of Freebase (see below). Freebase uses alphanumerical IDs as identifiers for entities (for example, *fb:m.05b6w* for Neil Armstrong) and human-readable strings with a directory-like structure as identifiers for predicates (for example, *fb:people.person.profession* for the predicate that links a person to their profession).[1] The dimensions of this dataset are: 1,934,771,338 triples, 32,281,516,161 word occurrences, 3,263,384,664 entity links.

Freebase was initiated by a company called Metaweb in 2007. The company was eventually acquired by Google in 2010. In August 2015, the Freebase dataset was frozen. Wikidata is a general-purpose knowledge base which is very similar in spirit to Freebase [15]. Wikidata has grown steadily but slowly until August 2017. It has then almost tripled in size over the next 12 months. A full dump of Wikidata from May 2018 contains 4,157,785,636 triples, of which 1,204,269,433 have a literal as an object. Wikidata uses numerical IDs both for entity and predicate names (for example, *wd:Q1615* for Neil Armstrong and *wdt:P106* for the predicate that links a person to their profession). An instance of the query engine described in Sect. 4 running on the complete Wikidata is available under http://qlever.cs.uni-freiburg.de.

3   Broccoli: Interactive Search on a Knowledge Base Combined with a Text Corpus

Search on a knowledge base alone is not easy, and in combination with text, the task becomes even harder. The main reason is that such a search requires knowledge of the names of the entities and the predicates in the knowledge base, as well as on how the information is structured in the knowledge base in the first place. This is hard even when the identifiers are human-readable (because there are so many of them, and names are ambiguous). It is complicated further when the identifiers are just numerical (or alphanumerical) IDs.

Another problem is that query languages like SPARQL have no concept of a ranking by *relevance*, as we know it from text search engines. Instead, a SPARQL query primarily delivers a *set* of entities or table rows, and any desired order needs to be explicitly specified. For some queries, there are natural predicates by which the data can be ordered. For example, for a list of cities, an order by descending population is natural. For other queries, there is no such predicate. For example, for a list of people, one probably wants to see the better known individuals first, but knowledge bases usually do not have predicates expressing the relative "popularity" of an entity.

Broccoli is a system that tries to address all of these problems. Broccoli guides the user in incrementally constructing a query by providing suggestions for extending the query after each keystroke and by visualizing, at each step of the construction process, the current query and the current result in an intuitive

---

[1] The identifiers are actually URIs and the prefix *fb:*stands for the common beginning of these URIs. See Sect. 5 for more explanation of this.

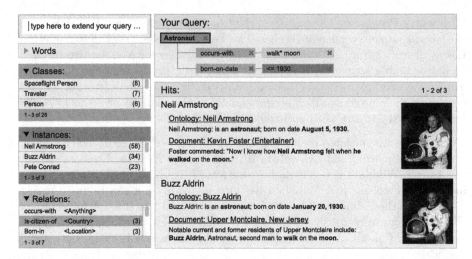

Fig. 1. A screenshot of Broccoli in action for an example query. The box on the top right visualizes the current query as a tree. The large box below shows the hits grouped by instances that match the query root and ranked by relevance. Comprehensive evidence for each hit is provided. For matches in the text corpus, a whole sentence is shown, with parts outside of the matching semantic unit (this is explained in the text) greyed out. On the left, there are suggestions for classes, instances and relations, which can be used to extend or narrow down the query. Suggested classes are parent classes of the entities from the current result. Relations and instances are context sensitive with respect to the current query. That is, all suggestions, if clicked, would lead to at least one hit. There are no word suggestions in the screenshot, because the search field on the top left is empty at this point of the query construction process. As soon as letters are typed, word suggestions appear and the other suggestions are narrowed down to those matching the typed prefix.

way. Figure 1 provides a screenshot of the system in action for a particular query. The caption of that figure provides some additional explanations on the various components and features. Note that Broccoli refers to entities as *instances* and predicates as *relations*. Broccoli also knows *classes*, which are simply groups of entities with the same type, according to a fixed type relation, which every general-purpose knowledge base has. The semantics of the query should be self-explanatory; if not, the formalization to SPARQL+Text in Sect. 4 should help to clarify this. A demo of Broccoli is available online at http://broccoli.cs.uni-freiburg.de.

The design and realization of Broccoli was a complex endeavour, which required several person-years. The architecture and the technology behind Broccoli is described in a series of papers. The system architecture has first been described in [1]. The index data structures and algorithms used for the interactive query processing and suggestions are described in [4]. The curation and simplification of the Freebase dataset is described in [2]. The engineering behind the public demo is described in [3]. The relevance scores which form the basis of the ranking are described in [6]. The natural language processing used to split the text into semantic units is described in [8].

Broccoli does what it does extremely well. The convenience and the high speed come at a price though. Here are the most important shortcomings of Broccoli:

1. Broccoli only supports tree-shaped SPARQL queries. Many typical queries are tree-shaped, but by far not all.
2. Broccoli only supports SPARQL queries with one variable in the SELECT clause. Again, many typical queries have this property, but by far not all.
3. Broccoli has no special treatment for n-ary relations, that is, relations which connect more than two entities. Section 5 gives an example of such a query on Wikidata.
4. Broccoli does not support SPARQL queries with predicate variables (that is, connecting two entities, which themselves may be variables, by an unknown predicate).
5. Broccoli has no query planner. Since the queries are constructed incrementally, by adding one part of a triple at a time, the order of the basic operations (index scans and joins, see Sect. 4), is completely determined by the way the query is constructed.
6. Broccoli uses a non-standard API. The original focus of the project was on usability aspects and efficiency, not on the underlying query language.

The query engine presented in the next section addresses all of these shortcomings.

4   QLever: A Query Engine for SPARQL+Text

QLever is a query engine for what we call SPARQL+Text. SPARQL+Text contains standard SPARQL, so QLever can also be used to process standard SPARQL queries (and quite efficiently so, see below). SPARQL+Text queries operate on a knowledge base linked to a text corpus, as explained in Sect. 2. It is assumed that the text has been segmented beforehand. These segments can be the semantic units of the sentences (as briefly explained in Sect. 3), or simply the sentences of the text. The search results are best if the segmentation is such that co-occurrence in the same segment has a semantic meaning, as in the "astronauts who walked on the moon" example above.[2]

Specifically, SPARQL+Text extends SPARQL by two built-in predicates *ql:contains-entity* and *ql:contains-word*. Here is an example query, which is similar to the query from Fig. 1 (but asking for the space agencies of the astronauts instead of restricting their birth date).

---

[2] We sweep under the rug here that this is not a matter of co-occurrence alone. For example, a text segment may additionally contain the word *not* and thus negate the meaning. There are different approaches to handle this which we do not discuss here.

SELECT ?astronaut ?agency TEXT(?text) **WHERE** {
   ?astronaut <is-a> <Astronaut>.
   ?astronaut <Space_Agency> ?agency.
   ?text ql:contains-entity ?astronaut.
   ?text ql:contains-word "walk*".
   ?text ql:contains-word "moon".
}
ORDER BY DESC(SCORE(?text))

The last three triples in the WHERE clause express that there is a segment of text, denoted by *?text*, which contains a mention of an entity *?astronaut* (which was identified as part of the NER+NED preprocessing described in Sect. 2) as well as a word starting with *walk* and the word *moon*. With a reasonable segmentation, a large number of such segments means that the respective candidate is indeed an astronaut who walked on the moon. On the two collections listed in Sect. 2, this query already gives very good results when the segments are simply sentences. The *TEXT(?text)* yields a third column in the result table, containing a matching text segment for each match for the remaining variables in the SELECT clause.[3] The final *ORDER BY ...* clause orders the results by the number of matching text segments, results with most matches first.

The details of the query engine behind QLever and the results of an extensive evaluation are provided in [5]. The source code and documentation is available under http://github.com/ad-freiburg/qlever. A front-end for entering SPARQL+Text queries on the datasets from Sect. 2 as well as on the complete Wikidata is available under http://qlever.cs.uni-freiburg.de.

We briefly provide the main ideas and results in a nutshell. The basic idea of the index structure behind QLever is similar to that of Broccoli, but with various engineering improvements. For example, the index data structures are laid out (with slight redundancy) such that all basic operations (such as obtaining a range of items from an index list) can efficiently read only exactly that data which is actually needed for the operations (for example, only the subjects from a range of triples). Due to these improvements, QLever is 2–3 times faster on queries which can also be processed by Broccoli.

Unlike Broccoli, QLever has a full-featured query planner, suited for arbitrary query graphs and with a special treatment of the two special *ql:contains-*... predicates. The query planner finds the query execution plan with the best estimated cost, using dynamic programming. For accurate cost estimations, for each subquery not only the result size is estimated, but also the number of distinct elements in each column. This is crucial for a good estimation of the result size of the join operations. Computing good estimates for the result size and the number of distinct elements in each column for each of QLever's basic operations (including the operations involving text search) is not trivial.

---

[3] The number of matching text segments shown (per match for the remaining variables in the SELECT clause) can be controlled with a *TEXTLIMIT <k>* clause. The default is TEXTLIMIT 1.

The performance improvements over existing SPARQL engines are considerable. Even for SPARQL queries without a text search component, QLever is several times faster than existing query engines like Virtuoso [14] (widely used in the commercial world) and RDF-3X [13] (a state-of-the-art research prototype) for most queries. On SPARQL+Text queries, which can only be simulated on query engines like Virtuoso and RDF-3X, QLever is faster by several orders of magnitude.

5   Outlook

QLever supports SPARQL+Text, a powerful extension of SPARQL to integrate search on a given text corpus linked to a given knowledge base. In particular, QLever addresses all the shortcomings of Broccoli listed at the end of Sect. 3. The price is that there is currently no UI that enables the construction of arbitrary SPARQL+Text queries with the same level of convenience as Broccoli.

To illustrate the need for such a UI, let us look at one more query. It is a SPARQL query on Wikidata for computing a table of all astronauts and where they studied and for which degree. This query involves a 3-ary predicate (linking a person to a university and the degree) and it is already surprisingly complex. What adds to the complexity is that Wikidata uses IDs for both entities and predicates and that we need to use a special predicate for obtaining the human-readable labels and additional FILTER clauses to get the labels only in one language instead of hundreds. Note that this query does not even involve a text search part (we could, however, just add the two text search triples from the query above). The prefixes at the beginning are abbreviations for the common prefixes of Wikidata's IDs, which are URIs. For example, in the SPARQL query below, *wd:Q11631* is equivalent to *<http://www.wikidata.org/entity/Q11631>*.

PREFIX wd: <http://www.wikidata.org/entity/>
PREFIX wdt: <http://www.wikidata.org/prop/direct/>
PREFIX p: <http://www.wikidata.org/prop/>
PREFIX pq: <http://www.wikidata.org/prop/qualifier/>
PREFIX ps: <http://www.wikidata.org/prop/statement/>
PREFIX rdfs: <http://www.w3.org/2000/01/rdf-schema#>
SELECT ?astronautLabel ?universityLabel ?degreeLabel
WHERE {
  ?astronaut wdt:P106 wd:Q11631.
  ?astronaut rdfs:label ?astronautLabel.
  ?astronaut p:P69 ?educatedAt.
  ?educatedAt ps:P69 ?university.
  ?university rdfs:label ?universityLabel.
  ?educatedAt pq:P512 ?degree.
  ?degree rdfs:label ?degreeLabel.
  FILTER langMatches(lang(?astronautLabel), "en")
  FILTER langMatches(lang(?universityLabel), "en")
  FILTER langMatches(lang(?degreeLabel), "en")
}

Enabling such queries, or more complex ones, without requiring knowledge of a particular query language or of the particularities of the knowledge base is a challenging problem. One line of attack is to try to generalize the UI of Broccoli to SPARQL+Text. This looks feasible, but it remains to be seen how practical such a UI would be when queries become more complex. Another line of attack is to automatically translate questions in natural language to SPARQL+Text queries as the above. This has been proven very hard already for simple SPARQL queries, for example see [9]. A compromise might be a hybrid approach, which allows an incremental construction of such a query via elements formulated in natural language.

References

1. Bast, H., Bäurle, F., Buchhold, B., Haussmann, E.: Broccoli: semantic full-text search at your fingertips. CoRR abs/1207.2615 (2012)
2. Bast, H., Bäurle, F., Buchhold, B., Haußmann, E.: Easy access to the freebase dataset. In: WWW (Companion Volume), pp. 95–98. ACM (2014)
3. Bast, H., Bäurle, F., Buchhold, B., Haußmann, E.: Semantic full-text search with broccoli. In: SIGIR, pp. 1265–1266. ACM (2014)
4. Bast, H., Buchhold, B.: An index for efficient semantic full-text search. In: CIKM, pp. 369–378. ACM (2013)
5. Bast, H., Buchhold, B.: QLever: a query engine for efficient SPARQL+text search. In: CIKM, pp. 647–656. ACM (2017)
6. Bast, H., Buchhold, B., Haussmann, E.: Relevance scores for triples from type-like relations. In: SIGIR, pp. 243–252. ACM (2015)
7. Bast, H., Buchhold, B., Haussmann, E.: Semantic search on text and knowledge bases. Found. Trends Inf. Retr. 10(2–3), 119–271 (2016)
8. Bast, H., Haussmann, E.: Open information extraction via contextual sentence decomposition. In: ICSC, pp. 154–159. IEEE Computer Society (2013)
9. Bast, H., Haussmann, E.: More accurate question answering on freebase. In: CIKM, pp. 1431–1440. ACM (2015)
10. Carmel, D., Chang, M., Gabrilovich, E., Hsu, B.P., Wang, K.: ERD'14: entity recognition and disambiguation challenge. In: SIGIR Forum, vol. 48, no. 2, pp. 63–77 (2014)
11. ClueWeb: The Lemur Projekt (2012). http://lemurproject.org/clueweb12
12. Cucerzan, S.: Large-scale named entity disambiguation based on Wikipedia data. In: EMNLP-CoNLL, pp. 708–716. ACL (2007)
13. Neumann, T., Weikum, G.: The RDF-3X engine for scalable management of RDF data. VLDB J. 19(1), 91–113 (2010). https://doi.org/10.1007/s00778-009-0165-y
14. OpenLink: The Virtuoso Project. https://virtuoso.openlinksw.com
15. Vrandecic, D., Krötzsch, M.: Wikidata: a free collaborative knowledgebase. Commun. ACM 57(10), 78–85 (2014)

A Tutorial on Query Answering and Reasoning over Probabilistic Knowledge Bases

İsmail İlkan Ceylan$^{(\boxtimes)}$ and Thomas Lukasiewicz$^{(\boxtimes)}$

University of Oxford, Oxford, UK
{ismail.ceylan,thomas.lukasiewicz}@cs.ox.ac.uk

Abstract. Large-scale probabilistic knowledge bases are becoming increasingly important in academia and industry alike. They are constantly extended with new data, powered by modern information extraction tools that associate probabilities with knowledge base facts. This tutorial is dedicated to give an understanding of various query answering and reasoning tasks that can be used to exploit the full potential of probabilistic knowledge bases. In the first part of the tutorial, we focus on (tuple-independent) probabilistic databases as the simplest probabilistic data model. In the second part of the tutorial, we move on to richer representations where the probabilistic database is extended with ontological knowledge. For each part, we review some known data complexity results as well as discuss some recent results.

Keywords: Probabilistic reasoning · Probabilistic databases
Probabilistic knowledge bases · Query answering · Data complexity

1 Introduction

In recent years, there has been a strong interest in building *large-scale probabilistic knowledge bases* from data in an automated way, which has resulted in a number of systems, such as DeepDive [65], NELL [50], Reverb [27], Yago [40], Microsoft's Probase [75], IBM's Watson [29], and Google's Knowledge Vault [25]. These systems continuously crawl the Web and extract *structured information*, and thus populate their databases with millions of entities and billions of tuples. To what extent can these search and extraction systems help with real-world use cases? This turns out to be an open-ended question. For example, Deep-Dive is used to build knowledge bases for domains such as paleontology, geology, medical genetics, and human movement [46,55]; it also serves as an important tool for the fight against human trafficking [35]. IBM's Watson makes impact on health-care systems [30] and many other application domains of life sciences.

This tutorial is mostly based on the dissertation work [13], and previously published material [8,14], and also makes use of some material from the classical literature on probabilistic databases [21,68].

© Springer Nature Switzerland AG 2018
C. d'Amato and M. Theobald (Eds.): Reasoning Web 2018, LNCS 11078, pp. 35–77, 2018.
https://doi.org/10.1007/978-3-030-00338-8_3

Google's Knowledge Vault has compiled more than a billion facts from the Web and is primarily used to improve the quality of search results on the Web [26], From a broader perspective, the quest for building large knowledge bases serves as a new dawn for artificial intelligence (AI) research in general and for unifying logic and probability in particular. Fields such as information extraction, natural language processing (e.g., question answering), relational and deep learning, knowledge representation and reasoning, and databases are taking initiative towards a common goal.

The most basic model to manage, store, and process large-scale probabilistic data is that of *tuple-independent probabilistic databases* (PDBs) [68], which indeed underlies many of these systems [25,65]. The first part of this tutorial is thus dedicated to give an overview on PDBs. We first recall the basics of query answering over PDBs, and then give an overview of the data complexity dichotomy between polynomial time and #P for evaluating unions of conjunctive queries over PDBs [21]. Then, we focus on two additional inference tasks that are inspired by *maximal posterior probability* computations in probabilistic graphical models (PGMs) [44]. That is, we discuss the problems of finding the *most probable database* and the *most probable hypothesis* for a given query, which intuitively correspond to finding explanations for PDB queries [14,37].

Probabilistic databases typically lack a suitable handling of incompleteness, in practice. In particular, each of the above systems encodes only a portion of the real-world, and this description is necessarily *incomplete*. However, when it comes to querying, most of these systems employ the *closed-world assumption* (CWA) [60], i.e., any fact that is not present in the knowledge base is assigned the probability 0, and thus assumed to be impossible. It is also common practice to view every extracted fact as an *independent* Bernoulli variable, i.e., any two facts are probabilistically independent. Similarly, by the *closed-domain assumption (CDA)* of PDBs, the domain of discourse is fixed to a *finite* set of *known* constants, i.e., it is assumed that all individuals are known a priori. These assumptions are very strong, and lead to more problematic semantic consequences, once combined with another limitation of these systems, namely, the lack of *commonsense knowledge*, which brings us to probabilistic knowledge bases.

In the second part of this tutorial, we focus on *probabilistic knowledge bases*, which are probabilistic databases that additionally allow to encode commonsense knowledge. Note that incorporating commonsense knowledge is inherently connected to giving up the above completeness assumptions of standard PDBs. In the scope of this tutorial, we assume that commonsense knowledge is encoded in the form of *ontologies*. There are many different models that allow for encoding commonsense knowledge; see e.g. [36], but most of these models, such as MLNs [61], relational Bayesian networks [42], and function-free variants of probabilistic logic programs employ the closed-domain assumption, and therefore, do not allow fully fledged first-order knowledge, as it already occurs in (rather restricted) ontology languages. These semantic differences have been recently highlighted in a survey [9].

Ontologies are first-order theories that formalize domain-specific knowledge, thereby allowing for *open-world, open-domain* reasoning. The most prominent ontology languages in the literature are based on description logics (DLs) [3], and lately also on Datalog$^\pm$ [10–12] (also studied as *existential rules*). For a uniform syntactic presentation, we focus on Datalog$^\pm$ ontologies, but we note that, in almost all cases, it is straight-forward to extend all the presented techniques and results. Interpreting databases under commonsense knowledge in the form of ontologies is closely related to ontology-based data access [56], also known as ontology-mediated query answering (OMQA) [7]. In the second part of this tutorial, we lift the reasoning problems introduced for probabilistic databases to ontology-mediated queries, and give an overview of various data complexity results.

2 Preliminaries: Logic, Databases, and Complexity

Intellectual roots of databases are in first-order logic [1]; in particular, in finite model theory [47]. Thus, we adopt the model-theoretic perspective and view databases as first-order structures over some fixed domain.

2.1 Logic and Notation

A *relational vocabulary* σ consists of sets $\mathbf{R}$ of *predicates*, $\mathbf{C}$ of *constants*, and $\mathbf{V}$ of *variable names* (or simply *variables*). The function $\mathrm{ar} : \mathbf{R} \mapsto \mathbb{N}$ associates with each predicate $\mathsf{P} \in \mathbf{R}$ a natural number, which defines the (unique) arity of P. A *term* is either a constant or a variable. An *atom* is of the form $\mathsf{P}(s_1, \ldots, s_n)$, where P is an n-ary predicate, and $s_1, \ldots, s_n$ are terms. A *ground atom* is an atom without variables.

A *first-order formula* is defined inductively by combining the logical atoms with logical connectives $\neg$, $\wedge$, $\vee$, and quantifiers $\exists$, $\forall$, as usual. A *literal* is either an atom or its negation. A *quantifier-free* formula is a formula that does not use a quantifier. A variable x in a formula Φ is *quantified*, or *bound*, if it is in the scope of a quantifier; otherwise, it is *free*. A *(first-order) sentence* is a first-order formula without any free variables, also called a *closed formula*, or *Boolean formula*. A *(first-order) theory* is a set of first-order sentences.

Let FO be the class of first-order formulas. The class of *existential first-order formulas* ($\exists$FO) consists of first-order formulas of the form $\exists \boldsymbol{x}.\Phi(\boldsymbol{x})$, where Φ is any Boolean combination of atoms. The class of *universal first-order formulas* ($\forall$FO) consists of first-order formulas of the form $\forall \boldsymbol{x}.\Phi(\boldsymbol{x})$, where Φ is any Boolean combination of atoms. A *disjunctive clause* is a finite disjunction of literals. A *conjunctive clause* is a finite conjunction of literals. The class of formulas in *existential conjunctive normal form* ($\exists$CNF) consists of first-order formulas of the form $\exists \boldsymbol{x}.\Phi(\boldsymbol{x})$; the class of formulas in *universal conjunctive normal form* ($\forall$CNF) consists of first-order formulas of the form $\forall \boldsymbol{x}.\Phi(\boldsymbol{x})$, where Φ is a conjunction of disjunctive clauses. The class of formulas in *existential disjunctive normal form* ($\exists$DNF) consists of formulas of the form $\exists \boldsymbol{x}.\Phi(\boldsymbol{x})$; the class of formulas in *universal disjunctive normal form* ($\forall$DNF) consists of formulas of the form $\forall \boldsymbol{x}.\Phi(\boldsymbol{x})$, where Φ is a disjunction of conjunctive clauses. The class of formulas in *conjunctive normal form* (CNF) consists of $\exists$CNF and $\forall$CNF formulas. The class of formulas in *disjunctive normal form* (DNF) consists of $\exists$DNF and $\forall$DNF formulas. A formula is *positive* if it contains only positive literals. We also write kCNF, or kDNF, to denote the class of formulas, where k denotes the maximal number of atoms that a clause can contain.

2.2 Databases and Query Answering

A *database* $\mathcal{D}$ over a (finite) relational vocabulary σ is a finite set of ground atoms over σ. We follow a model-theoretic approach and view a database as a *Herbrand interpretation*, where the atoms that appear in the database are mapped to *true*, while the ones not in the database are mapped to *false*,

Note that this semantics implies the *closed-world assumption* (CWA) [60], and the *closed-domain assumption* (CDA), i.e., restricts the domain of an interpretation to a *finite, fixed* set of constants; namely, to database constants. As a matter of fact, such interpretations presume that the *domain is complete*. Finally, the *unique name assumption* (UNA) ensures a bijection between the database constants and the domain: it is not possible to refer to the same individual in the domain with two different constant names. These simplifying assumptions of databases are useful for a variety of reasons. At the same time, it becomes very easy to produce some undesirable consequences under these assumptions, as we will elaborate. We will revisit some of these assumptions, and discuss their implications; for a recent survey on these assumptions, see e.g. [9].

The most fundamental task in databases is *query answering*; that is, given a database $\mathcal{D}$ and a formula $\Phi(x_1, \ldots, x_n)$ of first-order logic, to decide whether there exists assignments to the free variables $x_1, \ldots, x_n$ such that the resulting formula is satisfied by the database. Importantly, here, the variable assignments are of a special type, also called *substitutions*. Formally, a *substitution* $[x/t]$ replaces all occurrences of the variable x by some database constant t in some formula $\Phi[x, y]$, denoted $\Phi[x/t]$.

Given these, we can now formulate query answering as a decision problem. Let σ be a relational vocabulary; $\Phi(x_1, \ldots, x_n)$ be a first-order formula over σ; and $\mathcal{D}$ be a database over σ. Then, *query answering* is to decide whether $\mathcal{D} \models \Phi[x_1/a_1, \ldots, x_n/a_n]$ for a given substitution (answer) $[x_1/a_1, \ldots, x_n/a_n]$ to the free variables $x_1, \ldots, x_n$. For a Boolean formula Φ, *Boolean query answering* (or simply *query evaluation*) is to decide whether $\mathcal{D} \models \Phi$.

There exists a plethora of query languages in the literature. Classical database query languages range from the well-known *conjunctive queries* to arbitrary first-order queries, which we briefly introduce. A *conjunctive query* over σ is an existentially quantified formula $\exists \boldsymbol{x}.\Phi(\boldsymbol{x}, \boldsymbol{y})$, where $\Phi(\boldsymbol{x}, \boldsymbol{y})$ is a conjunction of atoms over σ. A *Boolean conjunctive query* over σ is a conjunctive query without free variables. A *union of conjunctive queries* is a disjunction of conjunctive queries with the same free variables. A *union of conjunctive queries* is Boolean if it does not contain any free variable. The class of *Boolean unions of conjunctive queries* is denoted UCQ.

We always focus on Boolean queries unless explicitly mentioned otherwise. Conjunctive queries are the most common database queries used in practice; thus, they will also be emphasized in this work. Besides, note that full relational algebra corresponds to the class of first-order formulas. Therefore, we include fragments of the class of first-order formulas as query languages in our analysis. In particular, we study ∃FO, ∀FO, and FO queries as query languages. Besides, we sometimes use different syntactic forms to represent relational queries, such as CNF or DNF.

We also speak of matches for Boolean queries. Let Q be a Boolean query over σ, $\mathcal{D}$ a database over σ, and $\mathbf{V}(Q)$ be the set of variables that occur in Q. A mapping $\varphi : \mathbf{V}(Q) \mapsto \mathbf{C}$ is called a *match for the query Q* in $\mathcal{D}$ if $\mathcal{D} \models \varphi(Q)$. For existentially quantified queries, it is sufficient to find a single match, to satisfy a given Boolean query evaluation. Conversely, for universally quantified queries, all mappings must result in a match in order to satisfy the query.

2.3 Complexity Background

We assume some familiarity with complexity theory and refer the reader to standard textbooks in the literature [53,66]. We now briefly introduce the complexity classes that are most relevant to the presented results.

FP is the class of functions $f : \{0,1\}^* \to \{0,1\}^*$ computable by a polynomial-time deterministic Turing Machine. The function class #P [71] is central for problems related to counting. The canonical problem for #P is #SAT, that is, given a propositional formula φ, the task of computing the number of satisfying assignments to φ. In this tutorial, we mostly focus on decision complexity classes. Intuitively, the complexity class PP [32] can be seen as the decision variant of #P. Formally, PP is the set of languages recognized by a polynomial time nondeterministic Turing machine that accepts an input if and only if *more than half* of the computation paths are accepting. The canonical problem for PP is MAJSAT, that is, given a propositional formula φ, the problem of deciding whether the majority of the assignments to φ are satisfying. Importantly, PP is closed under *truth table reductions* [6]; in particular, this implies that PP is closed under *complement, union,* and *intersection*. PP contains NP and is contained in PSPACE.

Another class of interest is $\mathrm{NP^{PP}}$, which intuitively combines search and optimization problems. A natural canonical problem for this class is EMAJSAT [48], that is, given a propositional formula φ and a set of distinguished variables $\boldsymbol{x}$ from φ, is there an assignment μ to $\boldsymbol{x}$-variables such that majority of the assignments τ that extend μ satifies φ. $\mathrm{NP^{PP}}$ appears as a fundamental class for probabilistic inference and planning tasks. The inclusion relationships between these complexity classes can be summarized as follows:

$$ \mathrm{P} \subseteq \mathrm{NP} \subseteq \mathrm{P^{PP}} = \mathrm{P^{\#P}} \subseteq \mathrm{NP^{PP}} \subseteq \mathrm{PSPACE} \subseteq \mathrm{EXP}, $$

where $\mathrm{P^{PP}} = \mathrm{P^{\#P}}$ is due to Toda's result [70].

We also make references to lower complexity classes characterized by Boolean circuits. AC^0 consists of the languages recognized by Boolean circuits with constant depth and polynomial number of *unbounded fan-in* AND and OR gates. TC^0 consists of the languages recognized by Boolean circuits with constant depth and a *polynomial number* of *unbounded fan-in* AND, OR, and MAJOR-ITY gates. Unlike Turing machines, Boolean circuits are non-uniform models of computation; that is, inputs of different size are processed by different circuits. It is therefore common to impose some *uniformity* conditions that require the existence of some resource-bounded Turing machine that, on an input, produces a description of the individual circuit. The most widely accepted uniformity condition for these classes is DLogTime-uniformity, bounding the computation in accordance to a logarithmic-time deterministic Turing machine. For a detailed treatment of the subject, we refer to the relevant literature [73]. The relationships between these classes can be summarized as follows:

$$AC^0 \subseteq TC^0 \subseteq \text{LogSpace} \subseteq P$$

When analyzing the computational complexity, we restrict ourselves to the data complexity throughout this tutorial, in order to achieve a more focused presentation. As usual, the *data complexity* is calculated only based on the size of the database, i.e., the query is assumed to be fixed [72].

3 Probabilistic Databases

In this section, we introduce probabilistic databases (PDBs) [68] as one of the most basic data models underlying probabilistic knowledge bases. Our analysis is organized in two subsections; first, we give an overview of *query evaluation* methods in PDBs, and afterwards, we study *maximal posterior computation* tasks, introduced for PDBs [14,37].

We adopt the simplest probabilistic database model, which is based on the *tuple-independence assumption*. For alternative models, we refer the reader to the rich literature of probabilistic databases; see e.g. [68] and the references therein. Tuple-independent probabilistic databases generalize classical databases by associating every database atom with a probability value.

Definition 1. A *probabilistic database* (PDB) $\mathcal{P}$ for a vocabulary σ is a finite set of *tuples* of the form $\langle t : p \rangle$, where t is a σ-atom and $p \in (0, 1]$. Moreover, if $\langle t : p \rangle \in \mathcal{P}$ and $\langle t : q \rangle \in \mathcal{P}$, then $p = q$.

Table 1 shows a sample PDB $\mathcal{P}_m$, where each row in a table represents an atom that is associated with a probability value. Semantically, a PDB can be viewed as a factored representation of exponentially many *possible worlds* (classical databases), each of which has a probability to be true. Both in the AI [24,57,63,64] and the database literature [68], this is commonly referred to as the *possible worlds semantics*.

In PDBs, each database atom is viewed as an independent random variable by the *tuple-independence assumption*. Each *world* is then simply a *classical*

Table 1. The PDB $\mathcal{P}_m$ represented in terms of database tables, where each row is interpreted as a probabilistic atom.

StarredIn		P
deNiro	taxiDriver	0.7
thurman	pulpFiction	0.1
travolta	pulpFiction	0.3

DirectedBy		P
pulpFiction	tarantino	0.8
taxiDriver	scorsese	0.6
winterSleep	ceylan	0.8

database, which sets a choice for all database atoms in the PDB. Furthermore, the *closed-world assumption* forces all atoms that are not present in the database to have probability zero.

Definition 2. A PDB $\mathcal{P}$ for vocabulary σ induces a *unique probability distribution* $\mathrm{P}_\mathcal{P}$ over the set of (possible worlds) $\mathcal{D}$ such that

$$\mathrm{P}_\mathcal{P}(\mathcal{D}) = \prod_{t \in \mathcal{D}} \mathrm{P}_\mathcal{P}(t) \prod_{t \notin \mathcal{D}} (1 - \mathrm{P}_\mathcal{P}(t)),$$

where the probability of each atom is given as

$$\mathrm{P}_\mathcal{P}(t) = \begin{cases} p \text{ if } \langle t : p \rangle \in \mathcal{P} \\ 0 \text{ otherwise.} \end{cases}$$

Whenever the probabilistic database is clear from the context, we simply write $\mathrm{P}(t)$, instead of $\mathrm{P}_\mathcal{P}(t)$. We say that a database is *induced by* a PDB $\mathcal{P}$ if it is a possible world (with a non-zero probability) of $\mathcal{P}$.

Observe that the choice of setting $\mathrm{P}_\mathcal{P}(t) = 0$ for tuples missing from PDB $\mathcal{P}$ is a *probabilistic counterpart* of the closed-world assumption. Let us now illustrate the semantics of PDBs on a simple example.

Example 1. Consider the PDB $\mathcal{P}_m$ given in Table 1. The probability of the world $\mathcal{D}_1$, given as

$$\mathcal{D}_1 := \{\mathsf{StarredIn}(\mathsf{deNiro}, \mathsf{taxiDriver}), \mathsf{DirectedBy}(\mathsf{taxiDriver}, \mathsf{scorsese})\},$$

can then be computed by multiplying the probabilities of the atoms that appear in $\mathcal{D}_1$ with the dual probability of the atoms that do not appear in $\mathcal{D}_1$ as follows

$$\mathrm{P}(\mathcal{D}_1) = 0.7 \cdot (1 - 0.1) \cdot (1 - 0.3) \cdot (1 - 0.8) \cdot 0.6 \cdot (1 - 0.8). \qquad \blacksquare$$

The semantics of queries is given through the possible worlds semantics, which amounts to walking through all the possible worlds and summing over the probabilities of those worlds that satisfy the query.

Definition 3 (query semantics). Let Q be a Boolean query and $\mathcal{P}$ be a PDB. The *probability* of Q in the PDB $\mathcal{P}$ is defined as

$$\mathrm{P}_\mathcal{P}(Q) = \sum_{\mathcal{D} \models Q} \mathrm{P}_\mathcal{P}(\mathcal{D}),$$

where $\mathcal{D}$ ranges over all possible worlds.

In general, there are exponentially many worlds, and in some cases, it is unavoidable to go through all of them in order to compute the probability. This is computationally intractable, but as we shall see later, in some cases, computing the query probability is actually easy.

Example 2. Consider again the PDB $\mathcal{P}_m$. In order to evaluate the following Boolean query

$$Q := \exists x, y \ \mathsf{StarredIn}(x, y) \wedge \mathsf{DirectedBy}(y, \mathsf{scorsese}),$$

on $\mathcal{P}_m$, we can naïvely check, for each world $\mathcal{D}$, whether $\mathcal{D} \models Q$. One such world is $\mathcal{D}_1$, as it clearly satisfies $\mathcal{D}_1 \models Q$. Afterwards, we only need to sum over the probabilities of the worlds, for which the satisfaction relation holds, in order to obtain the probability of the query.   ∎

In the given example, it is *easy* to compute the probability of the query. Notably, this is the case for any PDB and not only for our toy PDB. The next section is dedicated to give an understanding of *easy* and *hard* queries.

3.1   Query Evaluation in Probabilistic Databases

In this section, we provide a short overview on existing complexity results for inference in (tuple-independent) probabilistic databases including a data complexity dichotomy result. In our analysis, we are interested in the decision problem of probabilistic query evaluation, as defined next.

Definition 4 (probabilistic query evaluation). Given a PDB $\mathcal{P}$, a query Q and a threshold value $p \in [0, 1)$, *probabilistic query evaluation*, denoted PQE, is to decide whether $P_{\mathcal{P}}(Q) > p$. PQE is parametrized with a particular query language; thus, we write PQE(Q) to define PQE on the class of Q queries.

The data complexity of query evaluation depends heavily on the structure of the query. In a remarkable result [21], it has been shown that the probability of a UCQ can be computed either in FP or it is #P-hard on any PDB. Using the usual terminology [21], we say that queries are *safe* if the computation problem is in FP, and *unsafe*, otherwise. Probabilistic query evaluation, as defined here, is the corresponding decision problem. It is easy to see that this problem is either in P or it is PP-complete as a corollary to the result of [21].

Corollary 1. PQE(UCQ) *is either in* P *or it is* PP*-complete for PDBs in data complexity under polynomial-time Turing reductions.*

Proof. Let a UCQ Q be *safe* for PDBs. Then, for any PDB $\mathcal{P}$, the computation problem P(Q) uses only polynomial time. As a consequence, it is possible to decide whether the probability exceeds a given threshold p in polynomial time. Thus, PQE(UCQ) is in P for all *safe* UCQ queries.

Conversely, let a UCQ Q be *unsafe* for PDBs. Then, the problem of computing P(Q) on a given PDB $\mathcal{P}$ is #P-hard *under polynomial-time Turing reductions.*

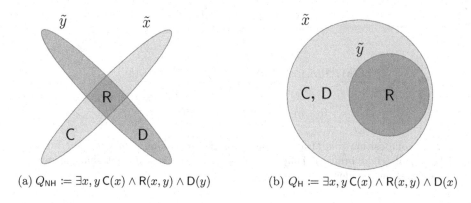

(a) $Q_{\mathsf{NH}} := \exists x, y\, \mathsf{C}(x) \wedge \mathsf{R}(x,y) \wedge \mathsf{D}(y)$ (b) $Q_{\mathsf{H}} := \exists x, y\, \mathsf{C}(x) \wedge \mathsf{R}(x,y) \wedge \mathsf{D}(x)$

Fig. 1. Venn diagram for the queries Q_{NH} (non-hierarchical) and Q_{H} (hierarchical).

Let us loosely denote by $P(Q)$ the problem of computing $P(Q)$. We now only show that PP is contained in $P^{\mathsf{PQE}(Q)}$, i.e., $\mathsf{PQE}(Q)$ is PP-hard in data complexity under polynomial-time Turing reductions.

To show this, let A be any other problem in PP. By assumption, its computation problem, denoted $\#\mathsf{A}$, is contained in $\mathrm{FP}^{P(Q)}$, i.e., there is a polynomial-time Turing machine with oracle $P(Q)$ that computes the output for $\#\mathsf{A}$. We can adapt this Turing machine then to compare the output to some threshold, which means that A is contained in $\mathrm{P}^{P(Q)}$. We also know that $P(Q)$ is contained in $\mathrm{FP}^{\mathsf{PQE}(Q)}$, as we can perform a binary search over the interval $[0, 1]$ to compute the precise probability $P(Q)$. This implies that A is contained in $\mathrm{P}^{\mathfrak{C}}$, where $\mathfrak{C} = \mathrm{FP}^{\mathsf{PQE}(Q)}$. Finally, note that the intermediate oracle does not provide any additional computational power (as this computation can be performed by the polynomial-time Turing machine and the oracle $\mathsf{PQE}(Q)$ can be queried directly). This shows that A is in $\mathrm{P}^{\mathsf{PQE}(Q)}$, which proves the result.   □

We use the same terminology also for the associated decision problem: we say that a query Q is *safe* if $\mathsf{PQE}(Q)$ is in P, and *unsafe*, otherwise. Historically, a dichotomy is considered first by [34], and later, the so-called *small dichotomy result* has been proven by [20], which applies to a subclass of conjunctive queries. As it gives nice insights on the larger dichotomy result [21], we look into this result, in more detail. The small dichotomy applies to all conjunctive queries without self-joins, i.e., conjunctive queries with non-repeating relation symbols. It asserts that a self-join free query is hard if and only if it is *nonhierarchical*, and it is safe, otherwise. It is therefore crucial to understand *hierarchical* and *nonhierarchical* queries.

Definition 5 (hierarchical queries). Let Q be a conjunctive query. For any variable x that appears in the query Q, its *x-cover*, denoted $\tilde{x}$, is defined as the set of all relation names that have the variable x as an argument. Two covers $\tilde{x}$

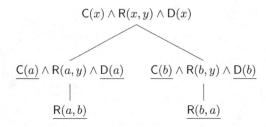

Fig. 2. Decomposition tree of a safe query for the grounding $[x/a, y/b]$ (left branch) and $[x/b, y/a]$ (right branch). Different branches of the tree do not share an atom, which ensures independence.

and $\tilde{y}$ are *pairwise hierarchical* if and only if $\tilde{x} \cap \tilde{y} \neq \emptyset$ implies $\tilde{x} \subseteq \tilde{y}$ or $\tilde{y} \subseteq \tilde{x}$. A query Q is *hierarchical* if every cover $\tilde{x}, \tilde{y}$ is *pairwise hierarchical*; otherwise, it is called *nonhierarchical*.

Let us consider the query $Q_{\mathsf{NH}} := \exists x, y\, \mathsf{C}(x) \wedge \mathsf{R}(x, y) \wedge \mathsf{D}(y)$. It is easy to see that this query is not hierarchical, since the relation R occurs in both covers $\tilde{x}$ and $\tilde{y}$ (as depicted in Fig. 1a). This simple join query is already unsafe, as shown in [20].

Theorem 1 (Theorem 7, [20]). Q_{NH} *is unsafe.*

Proof. We provide a reduction from the model counting problem, that is, given a propositional formula φ, the problem of computing the model count of φ, denoted $\#\varphi$. Provan and Ball [58] showed that computing $\#\varphi$ is $\#$P-complete (under Turing reductions) even for bipartite monotone 2DNF Boolean formulas φ, i.e., when the propositional variables can be partitioned into $X = x_1, \ldots, x_m$ and $Y = y_1, \ldots, y_n$ such that $\varphi = c_1 \vee \cdots \vee c_l$, where each clause c_i has the form $x_j \wedge y_k$, $x_j \in X, y_k \in Y$.

Given φ, we define the probabilistic database $\mathcal{P}_\varphi$, which contains the atoms $\langle \mathsf{C}(x_1) : 0.5 \rangle, \ldots, \langle \mathsf{C}(x_m) : 0.5 \rangle$, $\langle \mathsf{D}(y_1) : 0.5 \rangle, \ldots, \langle \mathsf{D}(y_n) : 0.5 \rangle$, and for every clause $(x_j \wedge y_k)$ an atom $\langle \mathsf{R}(x_j, y_k) : 1 \rangle$. It is then easy to verify that $\#\varphi = \mathsf{P}(Q_{\mathsf{NH}}) \cdot 2^{m+n}$, which concludes the result.   □

Similar reductions can be obtained for all (self-join free) conjunctive queries that are non-hierarchical. Note, however, that removing any of the atoms from Q_{NH} results in a safe query. For example, the query $\exists x, y\, \mathsf{C}(x) \wedge \mathsf{R}(x, y)$ is hierarchical and thus safe. The query $Q_{\mathsf{H}} := \exists x, y\, \mathsf{C}(x) \wedge \mathsf{R}(x, y) \wedge \mathsf{D}(x)$, as shown in Fig. 1b, is yet another example of a safe query.

The intuition behind a safe query is the query being recursively decomposable into sub-queries such that each such sub-query is probabilistically independent. Let us consider the query Q_{H}, as it admits a decomposition, and is safe. We can first ground over x, which results in a query of the form $\exists y\, \mathsf{C}(a) \wedge \mathsf{R}(a, y) \wedge \mathsf{D}(a)$ for a grounding $[x/a]$. The atoms in the resulting query do not share a relation name or a variable, and since we additionally assume tuple independence, it

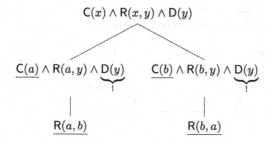

Fig. 3. Decomposition tree of an unsafe query for the grounding $[x/a, y/b]$ (left branch) and $[x/b, y/a]$ (right branch). Different branches of the tree share D-atoms, which makes them dependent.

follows that the probability of each atom is independent. Thus, their probabilities can be computed separately and combined afterwards using appropriate rules of probability.

Note that our observation for the independence is also valid for all different groundings of Q_H. For example, the groundings $Q_H[x/a]$ and $Q_H[x/b]$, are probabilistically independent, since after applying a grounding over y, we obtain *mutually disjoint sets* of ground atoms. That is, once x is mapped to different constants, then all mappings for y will result in different sets of atoms. As a result, their probabilities can be computed separately and combined afterwards. The decomposition of the safe query Q_H is depicted in Fig. 2 in terms of a tree. The key ingredient in this example is related to the variable x, which serves as a separator variable in the first place and allows us to further simplify the query.

Definition 6 (separator variable). Let Q be a first-order query. A variable x in Q is a *separator variable* if x appears in all atoms of Q and for any two different atoms of the same relation R, the variable x occurs in the same position.

Note that the query Q_{NH} has no separator variable, since neither x nor y serve as a separator variable. Intuitively, this means that the query cannot be decomposed into independent sub-queries. For example, two different groundings $Q_{NH}[x/a]$ and $Q_{NH}[x/b]$ are not independent for Q_{NH}, since they do not necessarily result in mutually exclusive sets of atoms once grounded over y, as shown in Fig. 3. The small dichotomy theorem uses other rules of probability theory to further simplify the query, and we illustrate this on an example.

Example 3. Consider the hierarchical query $Q_H := \exists x, y \; C(x) \wedge R(x, y) \wedge D(x)$. To compute the probability of Q_H, we first apply the decomposition based on the seperator variable x, which yields

$$P(Q_H) = 1 - \prod_{c \in C} P(\exists y \; C(c) \wedge R(c, y) \wedge D(c)),$$

Here, c ranges over the database constants, and the probability of the resulting expression can be computed by decomposing the conjunctions as

$$P(\exists y \; C(c) \wedge R(c, y) \wedge D(c)) = P(C(c)) \cdot P(\exists y \; R(c, y)) \cdot P(D(c)).$$

Table 2. The probabilistic database $\mathcal{P}_v$ encodes dietary regimes of some individuals, and the friendship relation among those individuals.

Vegetarian		FriendOf		Eats		Meat	
alice	0.7	alice bob	0.7	bob spinach	0.7	shrimp	0.7
bob	0.9	alice chris	0.8	chris mussels	0.8	mussels	0.9
chris	0.6	bob chris	0.1	alice broccoli	0.2	seahorse	0.3

The probabilities of the ground atoms $C(c)$, $D(c)$ can be read off from the given probabilistic database; thus, it only remains to apply the grounding in $R(c, y)$, which results in

$$P(\exists y\, R(c, y)) = 1 - \prod_{d \in C} P(R(c, d)).$$ ∎

The dichotomy for unions of conjunctive queries is much more intricate and a characterization of safe queries is unfortunately not easy. Thus, an algorithm is given in [21] to compute the probability of all safe queries by recursively applying the simplification rules on the query. This algorithm is complete, i.e., when the algorithm fails on the query, then the query is unsafe. Later, a lifted inference algorithm, called Lift^R, was proposed in [38], which was also proven to be complete. Afterwards, this algorithm has also been extended to an open-world semantics in [15], where it has also been noted that this algorithm runs in linear time for PDBs in the size of the database, under reasonable assumptions, such as unit arithmetic cost assumption for all arithmetic operations. For full details on different algorithms and their properties, we refer to the relevant literature [13,15,21,38].

3.2 Maximal Posterior Computations for Probabilistic Databases

Forming the foundations of large-scale knowledge bases, probabilistic databases have been widely studied in the literature. In particular, probabilistic query evaluation has been investigated intensively as a central inference mechanism. However, despite its power, query evaluation alone cannot extract all the relevant information encompassed in large-scale knowledge bases.

To exploit this potential, two additional inference tasks are proposed [14,37], namely, finding the *most probable database* and the *most probable hypothesis* for a given query, both of which are inspired by *maximal posterior probability* computations in probabilistic graphical models (PGMs) [44]. Let us briefly explain why probabilistic query evaluation alone may not be sufficient on the following example.

Example 4. Consider the PDB $\mathcal{P}_v$ given in Table 2 and the conjunctive query

$$Q_{\mathsf{fr}} = \exists x, y\, Q_{\mathsf{veg}}(x) \wedge \mathsf{FriendOf}(x, y) \wedge Q_{\mathsf{veg}}(y).$$

In the given PDB, alice, bob, and chris are all vegetarians and friends with each other with some probability. We can now ask the probability of vegetarians being friends with vegetarians. The query

$$Q_{\mathsf{fr}}[x/\mathsf{bob}] = \exists y \ Q_{\mathsf{veg}}(\mathsf{bob}) \wedge \mathsf{FriendOf}(\mathsf{bob}, y) \wedge Q_{\mathsf{veg}}(y)$$

is a special case of Q_{fr}, which asks whether bob has vegetarian friends. Its probability can be computed as $P(Q_{\mathsf{fr}}[x/\mathsf{bob}]) = 0.9 \cdot 0.1 \cdot 0.6 = 0.054$.

Suppose now that we observe that $Q_{\mathsf{fr}}[x/\mathsf{bob}]$ is true and would like to learn what best explains this observation relative to the underlying probabilistic database. To be able to explain such an observation, we need different inference tasks than probabilistic query answering.   ∎

The *most probable database* problem (analogous to *most probable explanations (MPE)* in PGMs), first proposed in [37], is the problem of determining the (classical) database with the largest probability that satisfies a given query. Intuitively, the query defines constraints on the data, and the goal is to find the most probable database that satisfies these constraints. The *most probable hypothesis* problem (analogous to *maximum a posteriori problems (MAP)* in PGMs), first proposed in [14], only asks for *partial* databases satisfying the query. The most probable hypothesis contains only atoms that contribute to the satisfaction condition of the query, which allows to more precisely pinpoint the most likely explanations of the query.

The Most Probable Database Problem. The need for alternative inference mechanisms for probabilistic knowledge bases has been observed before, and the *most probable database* problem has been proposed in [37] as follows.

Definition 7. Let $\mathcal{P}$ be a probabilistic database and Q a query. The *most probable database* for Q over $\mathcal{P}$ is given by

$$\arg\max_{\mathcal{D} \models Q} P(\mathcal{D}),$$

where $\mathcal{D}$ ranges over all worlds induced by $\mathcal{P}$.

Intuitively, a probabilistic database defines a probability distribution over exponentially many classical databases, and the most probable database is the element in this collection that has the highest probability, while still satisfying the given query. This can be seen as the best instantiation of a probabilistic model, and hence analogous to most probable explanation in Bayesian networks [23]. We illustrate the most probable database on our running example.

Example 5. Consider the given PDB $\mathcal{P}_v$. We can now filter the most probable database that satisfies the query

$$Q_{\mathsf{fr}} = \exists x, y \ Q_{\mathsf{veg}}(x) \wedge \mathsf{FriendOf}(x, y) \wedge Q_{\mathsf{veg}}(y).$$

Table 3. The most probable database for the query Q_{fr} over the PDB $\mathcal{P}_v$.

Vegetarian		FriendOf		Eats			Meat	
alice	0.7	alice bob	0.7	bob	spinach	0.7	shrimp	0.7
bob	0.9	alice chris	0.8	chris	mussels	0.8	mussels	0.9
chris	0.6	bob chris	0.1	alice	broccoli	0.2	seahorse	0.3

Intuitively, the atoms Vegetarian(alice), Vegetarian(bob), and FriendOf(alice, bob) need to be included in the most probable database, as they satisfy the query with the highest probability (compared to other possible matches). Since the query is monotone, for the remaining atoms, we only need to decide whether including them results in a higher probability than excluding them, which amounts to deciding whether their probability is greater than 0.5. Table 3 shows the most probable database for Q_{fr} over the PDB $\mathcal{P}_v$, where all atoms that belong to the most probable database are highlighted. ∎

Identifying the most probable database in this example is rather straightforward, and as we shall see later, this is tightly related to the query language that is considered. We now illustrate that identifying the most probable database can be more cumbersome if we consider other query languages; in particular, ∀FO queries.

Example 6. Consider again the PDB $\mathcal{P}_v$ and the query

$$Q_{veg} = \forall x, y\ \neg\mathsf{Vegetarian}(x) \vee \neg\mathsf{Eats}(x, y) \vee \neg\mathsf{Meat}(y),$$

which defines the constraints of being a vegetarian, which is violated by the atoms Vegetarian(chris), Eats(chris, mussels), and Meat(mussels). Therefore, the most probable database for Q_{veg} cannot contain all three of them, i.e., one of them has to be removed (from the explanation). In this case, it is easy to see that Vegetarian(chris) needs to be removed, as it has the lowest probability among them. Thus, the most probable database (in this case unique) contains all atoms of $\mathcal{P}_v$ that have a probability above 0.5, except for Vegetarian(chris).

Suppose now that we have observed $Q_{fr}[x/\mathsf{bob}]$, and we are interested in finding an explanation for this observation under the constraint of Q_{veg}, which is specified by the query

$$Q_{vf} = Q_{fr}[x/\mathsf{bob}] \wedge Q_{veg}.$$

Observe that Vegetarian(chris) and FriendOf(bob, chris) must be in the explanation to satisfy $Q_{fr}[x/\mathsf{bob}]$. Moreover, either Eats(chris, mussels) or Meat(mussels) has to be excluded from the most probable database, since otherwise Q_{veg} will be violated. The resulting most probable database is highlighted in Table 4. ∎

The most probable database is formulated as a decision problem as follows.

Table 4. The most probable database for the query Q_{vf} over the PDB $\mathcal{P}_v$.

Vegetarian		FriendOf		Eats		Meat	
alice	0.7	alice bob	0.7	bob spinach	0.7	shrimp	0.7
bob	0.9	alice chris	0.8	chris mussels	0.8	mussels	0.9
chris	0.6	bob chris	0.1	alice broccoli	0.2	seahorse	0.3

Definition 8 (MPD). Let Q be a query, $\mathcal{P}$ a probabilistic database, and $p \in (0, 1]$ a threshold. MPD is the problem of deciding whether there exists a database $\mathcal{D}$ that satisfies Q with $\mathrm{P}(\mathcal{D}) > p$. MPD is parametrized with a particular query language; thus, we write MPD(Q) to define MPD on the class Q of queries.

The first result that we present concerns the well-known class of *unions of conjunctive queries*: MPD can be solved using at most *logarithmic space* and *polynomial time*. More precisely, it is possible to encode the MPD problem uniformly into a class of TC^0 circuits.

Theorem 2. MPD(UCQ) *is in* DLOGTIME-*uniform* TC^0 *in data complexity.*

Proof. Let $\mathcal{P}$ be a PDB, Q a UCQ, and $p \in [0, 1)$ a threshold value. In principle, we can enumerate all the databases induced by the PDB $\mathcal{P}$ and decide whether there is a database that satisfies Q and has a probability greater or equal than p. This would require exponential time, as there are potentially exponentially many databases (worlds) induced by a probabilistic database. Fortunately, this can be avoided, as the satisfaction relation for unions of conjunctive queries is monotone: once a database satisfies a UCQ, then any superset of this database will satisfy the UCQ. We can, therefore, design an algorithm, which initiates the database with the atoms resulting from a match for the query (where the match is over the atoms that appear with a positive probability in the PDB $\mathcal{P}$) and extends this to a database with maximal probability, as follows: it adds only those atoms from the PDB $\mathcal{P}$ to the database that appear with a probability higher than 0.5, ensuring to obtain the database with the maximal probability. As a consequence, all these databases satisfy Q and if, furthermore, there is a database $\mathcal{D}$ among them with $\mathrm{P}(\mathcal{D}) \geq p$, then the algorithm answers yes; otherwise, it answers no.

It is easy to see that this algorithm is correct. However, if performed naïvely, as described, it results in a polynomial blow-up: there are polynomially many matches for the query (in data complexity), and for each match, constructing a database requires polynomial time and space. Observe, on the other hand, that we can uniformly access every match through an AC^0 circuit, and we can avoid explicitly constructing a database for each such match. More concretely, since we are only interested in the probability of the resulting database, we can perform an iterated multiplication over the probabilities of the individual atoms: if $\langle a : q \rangle \in \mathcal{P}$, where $q > 0.5$, we multiply with q, otherwise, we multiply with $1 - q$. We can perform such iterated multiplication with TC^0 circuits that can be constructed in DLOGTIME [39]. Finally, it is sufficient to compare the resulting number with the threshold value p. This last comparison operation can also be done using uniform AC^0 circuits. This puts $MPD(UCQ)$ in DLOGTIME-uniform TC^0 in data complexity.   □

This low data complexity result triggers an obvious question: what are the sources of tractability? The answer is partially hidden behind the fact that the satisfaction relation for unions of conjunctive queries is monotone. This allows us to reduce the search space to polynomially many databases, which are obtained from different matches of the query, in a unique way. Is the monotonicity a strict requirement for tractability? It turns out that the TC^0 upper bound can be strengthened towards all *existential queries*, i.e., existentially quantified formulas that allow negations in front of query atoms.

Theorem 3. $MPD(\exists FO)$ *is in* DLOGTIME-*uniform* TC^0 *in data complexity.*

Proof. Let $\mathcal{P}$ be a PDB, Q a $\exists FO$ query, and $p \in [0,1)$ a threshold value. Analogously to the proof of Theorem 2, we can design an algorithm that initiates the database with the positive atoms from a match, while now also keeping the record of the negative atoms from this match. We can again enumerate all the matches, and for each such match, perform an iterated multiplication over the probabilities of the atoms. The only difference is that now we need to exclude the atoms from the database that appear negatively in the match to ensure the satisfaction of the query. Thus, for all probabilistic atoms $\langle a : q \rangle \in \mathcal{P}$, we check whether $\neg a$ appears in the match, and if so, then we multiply with $1 - q$; otherwise, we check whether $q > 0.5$ and multiply with q, if this test is positive, and multiply with $1 - q$, if this test is negative. It is then sufficient to compare the result of each multiplication with the threshold value p: if one of the resulting multiplications is greater than or equal to p, then the algorithm answers yes, otherwise, it answers no. It is easy to verify the correctness of this algorithm.

By similar arguments as in the proof of Theorem 2, we can check the matches in AC^0, perform the respective multiplications in TC^0 and the respective comparisons in AC^0. Thus, $MPD(\exists FO)$ is in DLOGTIME-uniform TC^0 in data complexity.   □

In some sense, the presented tractability result implies that nonmonotonicity is not harmful if we restrict our attention to existential queries. This is because the nonmonotonicity involved here is of a limited type and does not lead to a combinatorial blow-up. This picture changes once we focus on universally quantified queries: nonmonotonicity combined with universal quantification creates nondeterministic choices, which we use to prove an NP-hardness result for the most probable database problem. We note that this result is very similar to the hardness result obtained in [37], only on a different query. Additionally, we show that the complexity bound is tight even if we consider FO queries.

Theorem 4. $MPD(\forall FO)$ *is* NP-*complete in data complexity, and so is* $MPD(FO)$.

Proof. We first show that MPD(FO) can be solved in NP. Let $\mathcal{P}$ be a PDB, Q a FO query, and $p \in [0, 1)$ a threshold value. To solve the decision problem, we first guess a world $\mathcal{D}$, and then verify both that the query is satisfied, i.e., $\mathcal{D} \models Q$, and that the threshold is met, i.e., $P(\mathcal{D}) \geq p$. Note that all these computations can be done using a nondeterministic Turing machine, since only the first step is nondeterministic, and both of the verification steps can be done in polynomial time in data complexity.

To prove hardness, we provide a reduction from the satisfiability of propositional 3CNF formulas. Let $\varphi = \bigwedge_i \varphi_i$ be a propositional formula in 3CNF. We define the ∀FO query

$$\begin{aligned} Q_{\mathsf{SAT}} := \forall x, y, z \ (\ &\mathsf{L}(x) \vee \ \mathsf{L}(y) \vee \ \mathsf{L}(z) \vee \mathsf{R}_1(x, y, z)) \wedge \\ &(\neg\mathsf{L}(x) \vee \ \mathsf{L}(y) \vee \ \mathsf{L}(z) \vee \mathsf{R}_2(x, y, z)) \wedge \\ &(\neg\mathsf{L}(x) \vee \neg\mathsf{L}(y) \vee \ \mathsf{L}(z) \vee \mathsf{R}_3(x, y, z)) \wedge \\ &(\neg\mathsf{L}(x) \vee \neg\mathsf{L}(y) \vee \neg\mathsf{L}(z) \vee \mathsf{R}_4(x, y, z)) \ , \end{aligned}$$

which is later used to encode the satisfaction conditions of φ. Without loss of generality, we denote with $u_1, \ldots, u_n$ the propositional variables that appear in φ.

We then define the PDB $\mathcal{P}_\varphi$, depending on φ, as follows. For each propositional variable u_j, we add the probabilistic atom $\langle \mathsf{L}(u_j) : 0.5 \rangle$ to the PDB $\mathcal{P}_\varphi$. The clauses φ_j are described with the help of the predicates $\mathsf{R}_1, \ldots, \mathsf{R}_4$, each of which corresponds to one type of clause. For example, if we have the clause $\varphi_i = x_1 \vee \neg x_2 \vee \neg x_4$, we add the atom $\langle \mathsf{R}_3(x_4, x_2, x_1) : 0 \rangle$ to $\mathcal{P}_\Phi$, which enforces via Q_{SAT} that either $\neg\mathsf{L}(x_4)$, or $\neg\mathsf{L}(x_2)$, or $\mathsf{L}(x_1)$ holds. All other R-atoms that do not correspond in such a way to one of the clauses are added with probability 1 to $\mathcal{P}_\Phi$.

The construction provided for Q_{SAT} and $\mathcal{P}_\varphi$ is clearly polynomial. Furthermore, the query is fixed, and only $\mathcal{P}_\varphi$ depends on φ. We now show that MPD can be used to answer the satisfiability problem of φ, using this construction.

Claim. The 3CNF formula φ is satisfiable if and only if there exists a database $\mathcal{D}$ induced by $\mathcal{P}_\varphi$ such that $P(\mathcal{D}) \geq (0.5)^n$ and $\mathcal{D} \models Q_{\mathsf{SAT}}$ (where n is the number of variables appearing in φ).

To prove the claim, suppose that φ is satisfiable, and let μ be such a satisfying assignment. We define a world $\mathcal{D}$ such that it contains all the atoms of the form $\mathsf{L}(u_j)$ if and only if $\mu(u_j) \mapsto 1$ in the given assignment. Moreover, $\mathcal{D}$ contains all the atoms that are assigned the probability 1 in $\mathcal{P}_\varphi$. It is easy to see that $\mathcal{D}$ is one of the worlds induced by $\mathcal{P}_\varphi$. Observe further that $\mathcal{P}_\varphi$ contains n nondeterministic atoms, each with 0.5 probability. By this argument, the probability of $\mathcal{D}$ is clearly $(0.5)^n$. It only remains to show that $\mathcal{D} \models Q_{\mathsf{SAT}}$, which is easy to verify.

For the other direction, let $\mathcal{D} \models Q_{\mathsf{SAT}}$ and $P(\mathcal{D}) \geq (0.5)^n$ for some world $\mathcal{D}$. We define an assignment μ by setting the truth value of u_j to 1, if $\mathsf{L}(u_j) \in \mathcal{D}$, and to 0, otherwise. Every world contains exactly one assignment for every variable, by our construction. Thus, the assignment μ is well-defined. It is easy to verify that $\mu \models \varphi$.   □

This concludes our analysis for the most probable database problem in the context of PDBs, and this problem is revisited later for ontology-mediated queries.

The Most Probable Hypothesis Problem. The most probable database identifies the most likely state of a probabilistic database relative to a query. However, it has certain limitations, which are analogous to the limitations of finding the *most probable explanations* in PGMs [44]. Most importantly, one is always forced to choose a *complete* database although the query usually affects only a subset of the atoms. That is, it is usually not the case that the whole database is responsible for the goal query to be satisfied. To be able to more precisely pinpoint the explanations of a query, the most probable hypothesis is introduced [14].

Definition 9. The *most probable hypothesis* for a query Q over a PDB $\mathcal{P}$ is

$$\arg\max_{\mathcal{H} \models Q} \sum_{\mathcal{D} \models \mathcal{H}} P(\mathcal{D}),$$

where $\mathcal{H}$ ranges over sets of atoms t and negated atoms $\neg t$ such that t occurs in $\mathcal{P}$, and $\mathcal{H} \models Q$ holds if and only if all worlds induced by $\mathcal{P}$ that satisfy $\mathcal{H}$ also satisfy Q.

Intuitively, the most probable hypothesis contains atoms only if they contribute to the satisfaction of the query, that is, the most probable hypothesis is a partial explanation. It is still the case that an explanation has to satisfy the query, but to do so, it does not need to make a decision for all the database atoms. Indeed, it is possible to specify some positive and negative atoms that ensure the satisfaction of the query, regardless of the truth value of the remaining database atoms.

Conversely, any database $\mathcal{D}$ with $\mathcal{D} \models \mathcal{H}$ must satisfy Q, and thus the most probable database can be seen as a special case of the most probable hypothesis that has to contain all atoms from a PDB (positively or negatively). We denote the sum inside the maximization, i.e., the probability of the explanation by $P(\mathcal{H})$. Differently from PGMs, the probability of the explanation can be computed by simply taking the product of the probabilities of the (negated) atoms in $\mathcal{H}$. This is a consequence of the independence assumption and influences the complexity results, as we elaborate later.

Example 7. Consider again our running example with the PDB $\mathcal{P}_v$ and recall the query $Q_{\mathsf{vf}} := Q_{\mathsf{veg}} \wedge Q_{\mathsf{fr}}[x/\mathsf{bob}]$, where

$$Q_{\mathsf{veg}} := \forall x, y \ \neg\mathsf{Vegetarian}(x) \vee \neg\mathsf{Eats}(x, y) \vee \neg\mathsf{Meat}(y),$$
$$Q_{\mathsf{fr}}[x/\mathsf{bob}] := \exists y \ Q_{\mathsf{veg}}(\mathsf{bob}) \wedge \mathsf{FriendOf}(\mathsf{bob}, y) \wedge Q_{\mathsf{veg}}(y).$$

Table 5. The most probable hypothesis for the query Q_{vf} over $\mathcal{P}_v$.

Vegetarian		FriendOf		Eats		Meat	
alice	0.7	alice bob	0.7	bob spinach	0.7	shrimp	0.7
bob	0.9	alice chris	0.8	chris mussels	0.8	mussels	0.9
chris	0.6	bob chris	0.1	alice broccoli	0.2	seahorse	0.3

Recall that the most probable database for Q_{vf} contains many redundant atoms. The most probable hypothesis $\mathcal{H}$ for the query Q_{vf} contains only 4 atoms, as given in Table 5: as before, light gray highlighting denotes positive atoms, while the dark gray one denotes negated atoms, i.e., ¬Eats(chris, mussels). Note that all of these atoms directly influence the satisfaction of the query and thus are part of the explanation. Since the most probable hypothesis contains less atoms, it is more informative than the most probable database. We can compute the probability of the hypothesis by $P(\mathcal{H}) = 0.9 \cdot 0.6 \cdot 0.1 \cdot (1 - 0.8) = 0.0108$. ∎

Whereas the most probable database represents full knowledge about all facts, which corresponds to the common closed-world assumption for (probabilistic) databases, the most probable hypothesis may leave atoms of $\mathcal{P}$ unresolved, which can be seen as a kind of open-world assumption (although the atoms that do not occur in $\mathcal{P}$ are still false). MPH is defined as a decision problem as follows.

Definition 10 (MPH). Let Q be a query, $\mathcal{P}$ a PDB, and $p \in (0, 1]$ a threshold. MPH is the problem of deciding whether there exists a hypothesis $\mathcal{H}$ that satisfies Q with $P(\mathcal{H}) \geq p$. MPH is parametrized with a particular query language; thus, we write MPH(Q) to define MPH on the class Q of queries.

We again start our analysis with *unions of conjunctive queries* and show that, as is the case for MPD, MPH can also be solved using at most *logarithmic space* and *polynomial time*.

Theorem 5. MPH(UCQ) *is in* DLogTime-*uniform* TC^0 *in data complexity.*

Proof. Let $\mathcal{P}$ be a PDB, Q a UCQ, and $p \in [0, 1)$ a threshold value. It has already been observed that the satisfaction relation for unions of conjunctive queries is monotone, i.e., the fact that once a database satisfies a UCQ, then any superset of this database satisfies the UCQ. Clearly, this also applies to partial explanations: the databases extending the hypothesis $\mathcal{H}$ satisfy the query only if $\mathcal{H}$ (extended with all atoms that have probability 1) is already a match for the query. This means that the hypothesis must be a subset of a ground instance of one of the disjuncts of the UCQ Q. The major difference from MPD is that, once the explanation is found, we do not need to consider the other database atoms in the PDB. As before, there are only polynomially many such hypotheses in the data complexity, and they can be encoded uniformly into AC^0 circuits, as in the proof of Theorem 2. Moreover, their probabilities can be computed in TC^0 and the comparison with the threshold p can be done again in AC^0. This puts MPH(UCQ) in DLogTime-uniform TC^0 in data complexity. □

Recall that, for MPD, it was possible to generalize the data tractability result from unions of conjunctive queries to existential queries. It is therefore interesting to know whether the same holds for MPH. Does MPH remain tractable if we consider existential queries? The answer is unfortunately negative: to be able to verify a test such as $\mathcal{H} \models Q$, we need to make sure that all extensions $\mathcal{D}$ of $\mathcal{H}$ satisfy the query and this test is hard, once we allow negations in front of query atoms. We illustrate the effect of negations on a simple example.

Example 8. Consider the following $\exists$FO query

$$Q := \exists x, y \ (A(x) \wedge \neg B(x,y)) \vee (\neg A(x) \wedge \neg B(x,y)).$$

To decide MPH on an arbitrary PDB, we could walk through all (partial) matches for the query and then verify $\mathcal{H} \models Q$. However, observe that this verification is not in polynomial time, in general, as there are interactions between the query atoms and these interactions need to be captured by the explanation itself. For instance, the A-atoms in Q are actually redundant, and it is enough to find an explanation that satisfies $\exists x, y \ \neg B(x,y)$ for some mapping. ∎

The next result is for existential queries, and via a reduction from the validity problem of 3DNF formulas, it shows that MPH is CONP-hard for these queries. It is easy to see that this is also a matching upper bound.

Theorem 6. MPH($\exists$FO) *is* CONP-*complete in data complexity.*

Proof. Let $\mathcal{P}$ be a PDB, Q an $\exists$FO query, and $p \in [0,1)$ a threshold value. As for membership, consider a nondeterministic Turing machine, which enumerates all partial matches forming the hypothesis $\mathcal{H}$ and answers yes if and only if $P(\mathcal{H}) \geq p$ and there is *no* database $\mathcal{D}$ that satisfies $\mathcal{D} \models \mathcal{H}$ while $\mathcal{D} \not\models Q$.

To prove hardness, we provide a reduction from the validity of propositional 3DNF formulas. Let $\varphi = \bigvee_i \varphi_i$ be a propositional formula in 3DNF. We first define the following $\exists$FO query

$$
\begin{aligned}
Q_{\mathsf{VAL}} := \exists x, y, z \ (\ & L(x) \wedge \ L(y) \wedge \ L(z) \wedge R_1(x,y,z)) \vee \\
& (\neg L(x) \wedge \ L(y) \wedge \ L(z) \wedge R_2(x,y,z)) \vee \\
& (\neg L(x) \wedge \neg L(y) \wedge \ L(z) \wedge R_3(x,y,z)) \vee \\
& (\neg L(x) \wedge \neg L(y) \wedge \neg L(z) \wedge R_4(x,y,z)) \, ,
\end{aligned}
$$

which is later used to encode the validity conditions of φ. Without loss of generality, let us denote with $u_1, \ldots, u_n$ the propositional variables that appear in φ. We then define the PDB $\mathcal{P}_\varphi$, depending on φ, as follows.

- For each propositional variable u_j, we add the probabilistic atom $\langle L(u_j) : 0.5 \rangle$ to the PDB $\mathcal{P}_\varphi$.
- The conjuncts φ_j are described with the help of the predicates $R_1, \ldots, R_4$, each of which corresponds to one type of conjunct. For example, if we have a clause $\varphi_j = u_1 \wedge \neg u_2 \wedge \neg u_4$, we add the atom $\langle R_3(u_4, u_2, u_1) : 1 \rangle$ to $\mathcal{P}_\Phi$, which enforces via Q_{VAL} that the conjunct that includes all atoms $\neg L(u_4)$, $\neg L(u_2)$, and $L(u_1)$ can be true. All other atoms $R_i(u_k, u_l, u_m)$ that do not correspond in such a way to one of the clauses are added with probability 0 to $\mathcal{P}_\Phi$.

The construction provided for Q_{VAL} and $\mathcal{P}_\varphi$ is clearly polynomial. Furthermore, the query is fixed, and only $\mathcal{P}_\varphi$ depends on φ. We now show that MPH can be used to answer the validity problem of φ, using this construction.

Claim. The 3DNF formula φ is valid if and only if there exists a hypothesis $\mathcal{H}$ over $\mathcal{P}_\Phi$ such that $\mathrm{P}(\mathcal{H}) \geq 1$ and $\mathcal{H} \models Q_{\mathsf{VAL}}$.

To prove the claim, suppose that φ is valid. We show that the empty hypothesis $\mathcal{H} = \emptyset$ satisfies both $\mathrm{P}(\mathcal{H}) \geq 1$ and $\mathcal{H} \models Q_{\mathsf{VAL}}$. It is easy to see that $\mathrm{P}(\mathcal{H}) = 1$ in $\mathcal{P}_\varphi$, as it encodes all possible databases, i.e., worlds. Let us assume by contradiction that there exists a database $\mathcal{D}$ that extends the hypothesis, $\mathcal{H} \models \mathcal{D}$, but does not satisfy the query, i.e., $\mathcal{D} \not\models Q_{\mathsf{VAL}}$. Then, we can use this database to define a valuation for the given propositional formula: define an assignment μ by setting the truth value of u_j to 1, if $\mathsf{L}(u_j) \in \mathcal{D}$, and to 0, otherwise. Every world contains exactly one assignment for every variable, by our construction. Thus, the assignment μ is well-defined. But then it is easy to see that this implies $\mu \not\models \varphi$, which contradicts the validity of φ.

For the other direction, let $\mathcal{H}$ be a hypothesis such that $\mathrm{P}(\mathcal{H}) \geq 1$ and $\mathcal{H} \models Q_{\mathsf{VAL}}$. This can only be the case if the $\mathcal{H}$ is the empty set (as otherwise $\mathrm{P}(\mathcal{H}) \leq 0.5$). This means that any database $\mathcal{D}$ induced by $\mathcal{P}_\varphi$ must satisfy the query. It is easy to see that every database is in one-to-one correspondence with a propositional assignment; thus, we conclude the validity of Q_{VAL}. $\qquad\square$

Having shown that MPH is harder than MPD for existential queries, one may wonder whether this is also the case for universal queries. We now show that MPH has the same complexity as MPD for universal queries.

Theorem 7. MPH($\forall$FO) *is NP-complete in data complexity.*

Proof. NP-hardness can be obtained analogously to the proof of Theorem 4, and we leave the hardness proof as an exercise, and focus on the upper bound which is not as straight-forward. Let $\mathcal{P}$ be a PDB, Q a $\forall$FO query, and $p \in [0, 1)$ a threshold value. To show membership, we nondeterministically guess a hypothesis $\mathcal{H}$ such that the satisfaction relation $\mathcal{H} \models Q$ is ensured. To do so, assume without loss of generality that the universal query is of the form

$$Q := \forall x_1, \dots, x_n \bigwedge_i q_i(x_1, \dots, x_n),$$

for some finite numbers $i, n > 0$, where each $q_i(x_1, \dots, x_n)$ is a clause over $x_1, \dots, x_n$. Our goal is to find a hypothesis that satisfies this query and that meets the threshold value p. To satisfy a universal query, we need to identify all database atoms that could possibly invalidate the query and rule them out. Thus, we consider the negation of the given query

$$\neg Q = \exists x_1, \dots, x_n \bigvee_i \neg q_i(x_1, \dots, x_n),$$

which encodes all database atoms that could possibly invalidate the original query Q. Furthermore, to make the effect of the database atoms more concrete, we consider all possible groundings of this query. Let us denote by

$$\neg Q[x_1/a_1, \ldots x_n/a_n]$$

a grounding with database constants a_i. There are polynomially many such groundings in data complexity. We need to ensure that none of the clauses in any of the groundings is satisfied by the hypothesis. Note that this can be achieved by including, for every clause, an atom into the hypothesis that contradicts the respective clause. For example, suppose that a grounding of the query is

$$(\mathsf{A}(a) \wedge \neg\mathsf{B}(b)) \vee \ldots \vee (\neg\mathsf{C}(c) \wedge \neg\mathsf{D}(d)).$$

Then, for the first clause, we have to either add the atom $\neg\mathsf{A}(a)$ or the atom $\mathsf{B}(b)$ to the hypothesis, and similarly for the last clause. The subtlety is that we cannot deterministically decide which atoms to include into the hypothesis (as there are interactions across clauses as well as across different groundings). Therefore, we nondeterministically guess each such choice. In essence, while constructing the hypothesis, we rule out everything that could invalidate the original query. As a consequence, we ensure that $\mathcal{H} \models Q$. It only remains to check whether $P(\mathcal{H}) \geq p$, which can be done in polynomial time. Thus, we obtain an NP upper bound in data complexity. □

MPH is coNP-complete for $\exists$FO queries (by Theorem 6) and NP-complete for $\forall$FO queries (by Theorem 7). By considering a particular query, which combines the power of existential and universal queries, it becomes possible to show Σ_2^{P}-hardness for MPH.

Theorem 8. MPH(FO) *is Σ_2^{P}-complete in data complexity.*

Proof. Let $\mathcal{P}$ be a PDB, Q a FO query, and $p \in [0,1)$ a threshold value. Consider a nondeterministic Turing machine with a (co)NP oracle: given a PDB $\mathcal{P}$, a first-order query Q, and a threshold $p \in (0,1]$, we can decide whether there exists a hypothesis $\mathcal{H}$ such that $P(\mathcal{H}) \geq p$ by first guessing a hypothesis $\mathcal{H}$, and verifying whether (i) $P(\mathcal{H}) \geq p$ and (ii) for *all* databases $\mathcal{D}$ that extend $\mathcal{H}$ and are induced by $\mathcal{P}$, it holds that $\mathcal{D} \models Q$. Verification of (i) can be done in deterministic polynomial time, and (ii) can be done in coNP (the complement is equivalent to the existence of an extension $\mathcal{D}$ and a valuation for the query variables that falsifies Q). This shows that MPH(FO) is in Σ_2^{P} in data complexity.

As for hardness, we provide a reduction from validity of quantified Boolean formulas of the form $\Phi = \exists u_1, \ldots, u_n \forall v_1, \ldots, v_m \, \varphi$, where φ is in 3DNF. Checking validity of such formulas is known to be Σ_2^{P}-complete. For the reduction, we consider the following query $Q = Q_{\mathsf{VAL}} \wedge Q_s$, where Q_{VAL} is the $\exists$FO query from Theorem 6 that encodes the validity conditions, and Q_s is a $\forall$FO query defined as

$$Q_s := \forall x \, (\neg\mathsf{E}(x) \wedge \mathsf{L}(x)) \vee (\mathsf{E}(x) \wedge \neg\mathsf{L}(x)) \vee \mathsf{F}(x).$$

Moreover, we define a PDB $\mathcal{P}_\Phi$ such that

- for each variable u that appears in φ, $\mathcal{P}_\Phi$ contains the atom $\langle \mathsf{L}(u) : 0.5 \rangle$;
- for each existentially quantified variable u_j, $\mathcal{P}_\Phi$ has the atom $\langle \mathsf{E}(u_j) : 0.5 \rangle$;
- for every universally quantified variable v_j, $\mathcal{P}_\Phi$ contains the atom $\langle \mathsf{F}(v_j) : 1 \rangle$;
- every conjunction in φ is described with the help of the predicates $\mathsf{R}_1, \ldots, \mathsf{R}_4$, each of which corresponds to one type of conjunctive clause. For example, if we have $\varphi_j = u_1 \wedge \neg u_2 \wedge \neg u_4$, we add the atom $\langle \mathsf{R}_3(u_4, u_2, u_1) : 1 \rangle$ to $\mathcal{P}_\Phi$, which enforces via Q_{VAL} that the clause that includes all atoms $\neg\mathsf{L}(u_4)$, $\neg\mathsf{L}(u_2)$, and $\mathsf{L}(u_1)$ can be true. Moreover, all the remaining R_i-atoms have probability 0.

In this construction, Q_{VAL} encodes the 3DNF, and Q_s helps us to distinguish between the existentially and universally quantified variables through the E- and F-atoms.

Claim. The quantified Boolean formula Φ is valid if and only if there exists a hypothesis $\mathcal{H}$ over $\mathcal{P}_\Phi$ such that $\mathrm{P}(\mathcal{H}) \geq (0.5)^{2n}$ and $\mathcal{H} \models Q$.

Suppose that Φ is valid. Then, there exists a valuation μ of $u_1, \ldots, u_n$, such that all valuations τ that extend this partial valuation (by assigning truth values to $v_1, \ldots, v_m$) satisfy φ. We define a hypothesis $\mathcal{H}$ depending on μ as follows. For all assignments $u_j \mapsto 1$ in μ, we add $\mathsf{L}(u_j)$ to $\mathcal{H}$; if, on the other hand, $u_j \mapsto 0$ in μ, we add $\neg\mathsf{L}(u_j)$ to $\mathcal{H}$. Moreover, to satisfy the query Q_s, for every $\mathsf{L}(u_j) \in \mathcal{H}$, we add $\neg\mathsf{E}(u_j)$ to $\mathcal{H}$, and analogously, for every $\neg\mathsf{L}(u_j) \in \mathcal{H}$, we add $\mathsf{E}(u_j)$ to $\mathcal{H}$. By this construction, there are clearly $2n$ atoms in $\mathcal{H}$, each of which has the probability 0.5 in $\mathcal{P}_\varphi$. Hence, it holds that $\mathrm{P}(\mathcal{H}) = (0.5)^{2n}$. Finally, it is sufficient to observe that all databases $\mathcal{D}$ that extend $\mathcal{H}$ must satisfy the query Q, as every such database is in one-to-one correspondence with a valuation τ that extends μ.

For the other direction, we assume that there exists a hypothesis $\mathcal{H}$ over $\mathcal{P}_\Phi$ such that $\mathrm{P}(\mathcal{H}) \geq (0.5)^{2n}$ and $\mathcal{H} \models Q$. This implies that $\mathcal{H}$ contains at most $2n$ atoms that have probability 0.5 in $\mathcal{P}_\varphi$ (and possibly some deterministic atoms). Furthermore, since $\mathcal{H} \models Q_s$, we know that $\mathcal{H}$ contains each E-atom either positively or negatively, and it also contains the complementary L-atom. Since these are already $2n$ atoms, $\mathcal{H}$ cannot contain any L-atoms for the universally quantified variables v_j. We can thus define a valuation μ for $u_1, \ldots, u_n$ simply by setting $u_j \mapsto 1$, if $\mathsf{L}(u_j) \in \mathcal{H}$, and $u_j \mapsto 0$, if $\neg\mathsf{L}(u_j) \in \mathcal{H}$. It is easy to see that the extensions τ of μ are in one-to-one correspondence with the databases that extend $\mathcal{H}$, and that φ evaluates to true for all of these assignments. $\qquad\square$

With this result, we conclude the data complexity analysis for MPH. We first showed a data tractability result concerning unions of conjunctive queries analogous to MPD. As before, this is the only result with no matching lower bound. Unlike MPD, however, $\exists$FO queries are proven to be coNP-complete for MPH in data complexity. Besides, MPH remains NP-complete for $\forall$FO queries, but turns out to be Σ_2^{P}-complete for first-order queries in data complexity.

4   Probabilistic Knowledge Bases

We have discussed several limitations of PDBs in the introduction, and stated that the unrealistic assumptions employed in PDBs lead to even more undesired consequences once combined with another limitation of these systems, namely, the lack of *commonsense knowledge*, a natural component of human reasoning, which is not present in plain (probabilistic) databases. A common way of encoding commonsense knowledge is in the form of ontologies.

In this section, we enrich probabilistic databases with ontological knowledge. More precisely, we adopt the terminology from [7] and speak of ontology-mediated queries (OMQs), that is, database queries (typically, unions of conjunctive queries) coupled with an ontology. The task of evaluating such queries is then called *ontology-mediated query answering (OMQA)*. We first give a brief overview on the Datalog$^\pm$ family of languages, and introduce the paradigm of ontology-mediated query answering. Afterwards, we present results regarding ontology-mediated query evaluation over PDBs. We also discuss maximal posterior reasoning problems in the context of ontology-mediated queries.

4.1   Ontology-Mediated Query Answering in Datalog$^\pm$

We again consider a relational vocabulary σ, which is now extended with a (potentially infinite) set $\mathbf{N}$ of nulls. We first introduce the so-called *negative constraints (NCs)*. From a database perspective, NCs can be seen as a special case of denial constraints over databases [67]. Formally, a *negative constraint (NC)* is a first-order formula of the form $\forall \boldsymbol{x}\, \Phi(\boldsymbol{x}) \to \bot$, where $\Phi(\boldsymbol{x})$ is a conjunction of atoms, called the *body* of the NC, and $\bot$ is the truth constant *false*. Consider, for example, the NCs

$$\forall x\, \mathsf{Writer}(x) \wedge \mathsf{Novel}(x) \to \bot,$$

$$\forall x, y\, \mathsf{ParentOf}(x, y) \wedge \mathsf{ParentOf}(y, x) \to \bot.$$

The former states that *writers* and *novels* are disjoint entities, whereas the latter asserts that the ParentOf relation is antisymmetric.

To formulate more general ontological knowledge, tuple-generating dependencies are introduced. Intuitively, such dependencies describe constraints on databases in the form of generalized Datalog rules with existentially quantified conjunctions of atoms in rule heads. Formally, a *tuple-generating dependency (TGD)* is a first-order formula of the form $\forall \boldsymbol{x}\, \Phi(\boldsymbol{x}) \to \exists \boldsymbol{y}\, \Psi(\boldsymbol{x}, \boldsymbol{y})$, where $\Phi(\boldsymbol{x})$ is a conjunction of atoms, called the *body* of the TGD, and $\Psi(\boldsymbol{x}, \boldsymbol{y})$ is a conjunction of atoms, called the *head* of the TGD. Consider the TGDs

$$\forall x, y\, \mathsf{AuthorOf}(x, y) \wedge \mathsf{Novel}(y) \to \mathsf{Writer}(x), \tag{1}$$

$$\forall y\, \mathsf{Novel}(y) \to \exists x\, \mathsf{AuthorOf}(x, y) \wedge \mathsf{Writer}(x). \tag{2}$$

The first one states that anyone who authors a novel is a writer. The second one asserts that all novels are authored by a writer. Note that TGDs can express the well-known inclusion dependencies and join dependencies from database theory. A *Datalog$^\pm$ program* (or *ontology*) Σ is a finite set of negative constraints and tuple generating dependencies. A Datalog$^\pm$ program is *positive* if it consists of only TGDs, i.e., does not contain any NCs.

In essence, Datalog$^\pm$ languages are only syntactic fragments of first-order logic, which also employ the *standard name assumption*, as in databases. Thus, a first-order interpretation $\mathcal{I}$ is a model of an ontology Σ in the classical sense, i.e., if $\mathcal{I} \models \alpha$ for all $\alpha \in \Sigma$. Given a database $\mathcal{D}$ defined over known constants and an ontology Σ, we write $mods(\Sigma, \mathcal{D})$ to represent the set of models of Σ that extend $\mathcal{D}$, which is formally defined as $\{\mathcal{I} \mid \mathcal{I} \models \mathcal{D}, \mathcal{I} \models \Sigma\}$. A database $\mathcal{D}$ is *consistent* w.r.t. Σ if $mods(\Sigma, \mathcal{D})$ is non-empty.

The entailment problem in Datalog$^\pm$ ontologies is undecidable [5], which motivated syntactic restrictions on Datalog$^\pm$ ontologies; there are a plethora of classes of TGDs [4,10,11,28,45]; here, we only focus on some of these classes. Our choice is primarily for ease of presentation, and most of the results can be extended to other classes in a straight-forward manner; see e.g. [13], for a more detailed analysis.

The (syntactic) restrictions on TGDs that we recall are guardedness [10], stickiness [12], and acyclicity, along with their "weak" counterparts, weak guardedness [10], weak stickiness [12], and weak acyclicity [28], respectively. A TGD is *guarded*, if there exists a body atom that contains (or "guards") all body variables. The class of guarded TGDs, denoted G, is defined as the family of all possible sets of guarded TGDs. A key subclass of guarded TGDs are the *linear* TGDs with just one body atom, which is automatically the guard. The class of linear TGDs is denoted by L. *Weakly guarded* TGDs extend guarded TGDs by requiring only the body variables that are considered "harmful" to appear in the guard (see [10] for full details). The associated class of TGDs is denoted WG. It is easy to verify that $\mathsf{L} \subset \mathsf{G} \subset \mathsf{WG}$.

Stickiness is inherently different from guardedness, and its central property can be described as follows: variables that appear more than once in a body (i.e., join variables) must always be propagated (or "stuck") to the inferred atoms. A TGD that enjoys this property is called *sticky*, and the class of sticky TGDs is denoted by S. Weak stickiness generalizes stickiness by considering only "harmful" variables, and defines the class WS of *weakly sticky* TGDs. Observe that $\mathsf{S} \subset \mathsf{WS}$.

A set of TGDs is *acyclic* and belongs to the class A if its predicate graph is acyclic. Equivalently, an acyclic set of TGDs can be seen as a non-recursive set of TGDs. A set of TGDs is *weakly acyclic*, if its dependency graph enjoys a certain acyclicity condition, which guarantees the existence of a finite canonical model; the associated class is denoted WA. Clearly, $\mathsf{A} \subset \mathsf{WA}$. Interestingly, it also holds that $\mathsf{WA} \subset \mathsf{WS}$ [12].

Another fragment of TGDs are *full* TGDs, i.e., TGDs without existentially quantified variables. The corresponding class is denoted by F. Restricting full TGDs to satisfy linearity, guardedness, stickiness, or acyclicity yields the classes LF, GF, SF, and AF, respectively. It is known that $\mathsf{F} \subset \mathsf{WA}$ [28] and $\mathsf{F} \subset \mathsf{WG}$ [10].

We usually omit the universal quantifiers in TGDs and NCs, and for clarity we consider single-atom-head TGDs; however, our results can be easily extended to TGDs with conjunctions of atoms in the head (except under the bounded-arity assumption). Following the common convention, we will assume that NCs are part of all Datalog$^\pm$ languages.

Table 6. A database that consists of the unary relations Writer, Novel and the binary relation AuthorOf.

Writer	AuthorOf		Novel
balzac	hamsun	hunger	goriot
dostoyevski	dostoyevski	gambler	hunger
kafka	kafka	trial	trial

Ontology-mediated query answering is a popular paradigm for querying incomplete data sources in a more adequate manner [7]. Formally, an *ontology-mediated query (OMQ)* is a pair $(Q, \mathcal{T})$, where Q is a Boolean query, and $\mathcal{T}$ is an ontology. Given a database $\mathcal{D}$ and an OMQ $(Q, \mathcal{T})$, we say that $\mathcal{D}$ entails the OMQ $(Q, \mathcal{T})$, denoted $\mathcal{D} \models (Q, \mathcal{T})$, if for all models $\mathcal{I} \models (\mathcal{T}, \mathcal{D})$ it holds that $\mathcal{I} \models Q$. Then, *ontology-mediated query answering (OMQA)* is the task of deciding whether $\mathcal{D} \models (Q, \mathcal{T})$ for a given database $\mathcal{D}$ and an OMQ $(Q, \mathcal{T})$. Note that we use the term query answering in a rather loose sense to refer to the Boolean query evaluation problem.

Example 9. Let us consider the database $\mathcal{D}_a$ given in Table 6. Observe that the simple queries

$$Q_1 := \mathsf{Writer(hamsun)} \quad and \quad Q_2 := \exists x\, \mathsf{Writer}(x) \wedge \mathsf{AuthorOf}(x, \mathsf{goriot})$$

are not satisfied by the database $\mathcal{D}_a$ although they should evaluate to true from an intuitive perspective. On the other side, under the Datalog$^\pm$ program Σ_a that consists of the TGDs

$$\forall x, y\, \mathsf{AuthorOf}(x, y) \wedge \mathsf{Novel}(y) \rightarrow \mathsf{Writer}(x),$$
$$\forall y\, \mathsf{Novel}(y) \rightarrow \exists x\, \mathsf{AuthorOf}(x, y) \wedge \mathsf{Writer}(x),$$

both of these queries are satisfied: $\mathcal{D}_a \models (Q_1, \Sigma_a)$ holds due to the first rule, and $\mathcal{D}_a \models (Q_2, \Sigma_a)$ holds due to the second rule. The incomplete database is queried through the logical rules that encode commonsense knowledge, which in turn results in more complete answers. ∎

A key paradigm in ontology-mediated query answering is the *first-order rewritability* of queries. Intuitively, FO-rewritability ensures that we can rewrite an OMQ into a (possibly large) UCQ, and this transformation is *homomorphism-preserving over all finite structures* [62]. More formally, let $\mathcal{T}$ be an ontology and Q a Boolean query. Then, the OMQ $(Q, \mathcal{T})$ is *FO-rewritable* if there exists a Boolean UCQ $Q_\mathcal{T}$ such that, for all databases $\mathcal{D}$ that are consistent w.r.t. $\mathcal{T}$, we have $\mathcal{D} \models (Q, \mathcal{T})$ if and only if $\mathcal{D} \models Q_\mathcal{T}$. In this case, $Q_\mathcal{T}$ is called an *FO-rewriting* of $(Q, \mathcal{T})$. A language $\mathcal{L}$ is *FO-rewritable* if it admits an FO-rewriting for any UCQ and theory in $\mathcal{L}$.

FO-rewritability implies a data-independent reduction from OMQA to query evaluation in relational databases. In practical terms, this means that the query

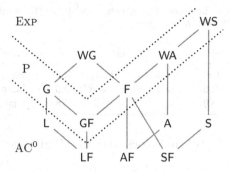

Fig. 4. Inclusion relationships and data complexity of ontology-mediated query answering for Datalog$^\pm$ languages.

can be rewritten into an SQL query to be evaluated in relational database management systems. In theoretical terms, this puts OMQA in AC^0 in data complexity for all FO-rewritable languages. The data complexity of ontology-mediated query answering for basic Datalog$^\pm$ languages is summarized in Fig. 4.

4.2   Ontology-Mediated Queries for Probabilistic Databases

We give an overview of the problem of evaluating ontology-mediated queries for PDBs. The idea is to allow Datalog$^\pm$ programs on top of tuple-independent PDBs and query the probability of a given OMQ.

Definition 11 (semantics). The *probability of an OMQ* (Q, Σ) *relative to a probability distribution* P is

$$P(Q, \Sigma) = \sum_{\mathcal{D} \models (Q, \Sigma)} P(\mathcal{D}),$$

where $\mathcal{D}$ ranges over all databases over σ.

The major difference compared to PDBs is that this semantics defers the decision of whether a world satisfies a query to an entailment test, which also includes a logical theory.

Note that the Datalog$^\pm$ program can be inconsistent with some of the worlds, which makes standard reasoning very problematic, as *anything* can be entailed from an inconsistent theory (*"ex falso quodlibet"*). The common way of tackling this problem in probabilistic knowledge bases is to restrict probabilistic query evaluation to only consider consistent worlds by setting the probabilities of inconsistent worlds to 0 and *renormalizing* the probability distribution over the set of worlds accordingly. More formally, the query probabilities can be normalized by defining

$$P_n(Q, \Sigma) := (P(Q, \Sigma) - \gamma)/(1 - \gamma),$$

Table 7. A probabilistic database $\mathcal{P}_a$, which consists of the unary relations Writer, Novel and the binary relation AuthorOf.

Writer	P		AuthorOf		P		Novel	P
balzac	0.8		hamsun	hunger	0.9		goriot	0.7
dostoyevski	0.6		dostoyevski	gambler	0.6		hunger	0.4
kafka	0.9		kafka	trial	0.8		trial	0.5

where γ is the probability of the inconsistent worlds given as

$$\gamma := \sum_{mods(\Sigma,\mathcal{D})=\emptyset} P(\mathcal{D}).$$

The normalization factor γ can thus be computed once and then reused as a post-processing step. Hence, for simplicity, we assume that all the worlds $\mathcal{D}$ induced by the PDB are consistent with the program.

Let us now illustrate the effect of ontological rules in PDBs. Recall Example 9, where we illustrated the effect of ontological rules in querying databases. Queries that intuitively follow from the knowledge encoded in the database were not satisfied by the given database (from Table 6). Still, it was possible to alleviate this problem using ontological rules. We now adopt this example to PDBs and observe a similar effect.

Example 10. Let us consider the PDB $\mathcal{P}_a$ given in Table 7. The queries

$$Q_1 := \mathsf{Writer(hamsun)} \quad and \quad Q_2 := \exists x\, \mathsf{Writer}(x) \wedge \mathsf{AuthorOf}(x, \mathsf{goriot})$$

from Example 9 evaluate to the probability 0 on $\mathcal{P}_a$. On the other side, under the Datalog$^{\pm}$ program Σ_a that consists of the rules

$$\forall x, y\, \mathsf{AuthorOf}(x, y) \wedge \mathsf{Novel}(y) \rightarrow \mathsf{Writer}(x),$$
$$\forall y\, \mathsf{Novel}(y) \rightarrow \exists x\, \mathsf{AuthorOf}(x, y) \wedge \mathsf{Writer}(x),$$

there are worlds where both of these queries are satisfied. One such world is $\mathcal{D}_a$ given in Table 6, i.e., recall that $\mathcal{D}_a \models (Q_1, \Sigma_a)$ and $\mathcal{D}_a \models (Q_2, \Sigma_a)$.

More precisely, we obtain that $P(Q_1) = 0.63$, since any world that contains both AuthorOf(hamsun, hunger) and Novel(hunger) entails Writer(hamsun), and no other world does. Similarly, $P(Q_2) = 0.7$, since Q_2 is entailed from all and only those worlds where Novel(goriot) holds. ∎

Probabilistic query evaluation, as a decision problem, is defined as before with the only difference that it is now parametrized with OMQs.

Definition 12 (probabilistic query evaluation). Given a PDB $\mathcal{P}$, an OMQ (Q, Σ), and a value $p \in [0, 1)$, *probabilistic query evaluation*, denoted PQE, is to decide whether $P_{\mathcal{P}}(Q, \Sigma) > p$ holds. PQE is parametrized with the language of the ontology and the query; we write PQE$(Q, \mathcal{L})$ to define PQE on the class Q of queries and on the class of ontologies restricted to the language $\mathcal{L}$.

We will deliberately use the terms probabilistic query evaluation and probabilistic OMQ evaluation interchangeably if there is no danger of ambiguity. We now provide a host of complexity results for probabilistic OMQ evaluation relative to different languages.

We start our complexity analysis with a rather simple result that is of a generic nature. Intuitively, given the complexity of OMQA in a Datalog$^\pm$ language, we obtain an immediate upper and lower bound for the complexity of probabilistic OMQ evaluation in that language.

Theorem 9. *Let $\mathfrak{C}$ denote the data complexity of ontology-mediated query answering for a Datalog$^\pm$ language $\mathcal{L}$. Then, $\mathsf{PQE}(\mathsf{UCQ}, \mathcal{L})$ is $\mathfrak{C}$-hard and in $\mathrm{PP}^{\mathfrak{C}}$ for PDBs in data complexity.*

Proof. Let (Q, Σ) be an OMQ, $\mathcal{P}$ be a PDB, and $p \in [0, 1)$ be a threshold value. Consider a nondeterministic Turing machine with a $\mathfrak{C}$ oracle. Each branch corresponds to a world $\mathcal{D}$ and is marked as either accepting or rejecting depending on the outcome of the logical entailment check $\mathcal{D} \models (Q, \Sigma)$. This logical entailment test is in $\mathfrak{C}$ for the language $\mathcal{L}$, by our assumption. Thus, it can be performed using the oracle. Then, by this construction, the nondeterministic Turing machine answers yes if and only if $\mathrm{P}_{\mathcal{P}}(Q, \Sigma) > p$, which proves membership to $\mathrm{PP}^{\mathfrak{C}}$ in the respective complexity.

To show $\mathfrak{C}$-hardness, we reduce from ontology-mediated query answering, that is, given a database $\mathcal{D}$ and an OMQ (Q, Σ), where Q is a UCQ, and Σ is a program over $\mathcal{L}$, decide whether $\mathcal{D} \models (Q, \Sigma)$. We define a PDB $\mathcal{P}$ that contains all the atoms from the database $\mathcal{D}$ with probability 1. Then, it is easy to see that $\mathcal{D} \models (Q, \Sigma)$ if and only if $\mathrm{P}_{\mathcal{P}}(Q) \geq 1$.   □

We briefly analyze the consequences of Theorem 9. Observe that, for all deterministic complexity classes $\mathfrak{C}$ that contain PP, it holds that $\mathrm{PP}^{\mathfrak{C}} = \mathfrak{C}$, and thus Theorem 9 directly implies tight complexity bounds. For instance, the data complexity of probabilistic OMQ evaluation for WG is Exp-complete as a simple consequence of Theorem 9.

Beyond this generic result, one is interested in lifting the data complexity dichotomy for unions of conjunctive queries to OMQs. This connection is immediate, as shown in the following.

Lemma 1. *Let (Q, Σ) be an OMQ, where Q is a UCQ, and Σ is a program, and Q_Σ be an FO-rewriting of (Q, Σ). Then, for any PDB $\mathcal{P}$, it holds that $\mathrm{P}_{\mathcal{P}}(Q, \Sigma) = \mathrm{P}_{\mathcal{P}}(Q_\Sigma)$.*

Proof. For any PDB $\mathcal{P}$, it holds that

$$\mathrm{P}_{\mathcal{P}}(Q, \Sigma) \overset{(1)}{=} \sum_{\mathcal{D} \models (Q, \Sigma)} \mathrm{P}_{\mathcal{P}}(\mathcal{D}) \overset{(2)}{=} \sum_{\mathcal{D} \models Q_\Sigma} \mathrm{P}_{\mathcal{P}}(\mathcal{D}) \overset{(3)}{=} \mathrm{P}_{\mathcal{P}}(Q_\Sigma),$$

where (1) follows from Definition 11; (2) follows from Q_Σ being the FO-rewriting of Q w.r.t. Σ; and (3) is the definition of the semantics of Q_Σ in PDBs.   □

With the help of Lemma 1, it becomes possible to lift the data complexity dichotomy in probabilistic databases to all Datalog$^\pm$ languages that are FO-rewritable.

Theorem 10 (dichotomy). *For all FO-rewritable Datalog$^\pm$ languages $\mathcal{L}$, the following holds. PQE(UCQ, $\mathcal{L}$) is either in P or it is PP-complete for PDBs in data complexity under polynomial-time Turing reductions.*

Proof. Let (Q, Σ) be an OMQ, where Q is a UCQ, and Σ is a Datalog$^\pm$ program over an FO-rewritable language, and Q_Σ be an FO-rewriting of (Q, Σ). By Lemma 1, any polynomial-time algorithm that can evaluate Q_Σ over PDBs also yields the probability of the OMQ (Q, Σ) relative to an PDB, and vice versa. This implies that the OMQ (Q, Σ) is safe if Q_Σ is safe.

Dually, by the same result, the probabilities of *all* rewritings of Q coincide, and hence the same algorithm can be used for all of them. Thus, if (Q, Σ) is unsafe, then Q_Σ must also be unsafe for PDBs. By the dichotomy of [21] and Lemma 1, this implies that evaluating the probability of both the UCQ Q_Σ and the OMQ (Q, Σ) must be PP-hard under Turing reductions. □

Obviously, ontological rules introduce dependencies. Therefore, a safe query can become unsafe for OMQs. However, the opposite effect is also possible, i.e., an unsafe query may become safe under ontological rules. We illustrate both of these effects on a synthetic example.

Example 11. Consider the conjunctive query $\exists x, y\, C(x) \wedge D(x, y)$, which is safe for PDBs. It becomes unsafe under the TGD $R(x, y), T(y) \to D(x, y)$, since then it rewrites to the query

$$(\exists x, y\, C(x) \wedge D(x, y)) \vee (\exists x, y\, C(x) \wedge R(x, y) \wedge T(y)),$$

which is unsafe. Conversely, the conjunctive query $\exists x, y\, C(x) \wedge R(x, y) \wedge D(y)$ is not safe for PDBs, but becomes safe under the TGD $R(x, y) \to D(y)$, as it rewrites to $\exists x, y\, C(x) \wedge R(x, y)$. Note that these are very simple TGDs, which are full, acyclic, guarded, and sticky. ∎

Recall that the PP-hardness of probabilistic UCQ evaluation in data complexity holds under polynomial-time Turing reductions. This transfers to probabilistic OMQ evaluation relative to FO-rewritable languages. On the other hand, for guarded full programs, it is possible to show that PP-hardness holds even under standard many-one reductions.

Theorem 11. PQE(UCQ, GF) *is PP-hard for PDBs in data complexity.*

Proof. We reduce the following problem [74]: decide the validity of the formula $\Phi = \mathsf{C}^c x_1, \ldots, x_n\, \varphi$, where $\varphi = \varphi_1 \wedge \cdots \wedge \varphi_k$ is a propositional formula in CNF over the variables $x_1, \ldots, x_n$. This amounts to checking whether there are at least c assignments to $x_1, \ldots, x_n$ that satisfy φ. We assume without loss of generality that φ contains all clauses of the form $x_j \vee \neg x_j$, $1 \le j \le n$; clearly, this does not affect the existence or number of satisfying assignments for φ. For the reduction, we use a PDB $\mathcal{P}_\Phi$ and a program Σ_Φ. We first define the PDB $\mathcal{P}_\Phi$ as follows.

- For each variable x_j, $1 \leq j \leq n$, $\mathcal{P}_\Phi$ contains the probabilistic atoms $\langle \mathsf{L}(x_j, 0) : 0.5 \rangle$ and $\langle \mathsf{L}(x_j, 1) : 0.5 \rangle$, where we view x_j as a constant. These atoms represent the assignments that map x_j to *false* and *true*, respectively.
- For each propositional literal $(\neg)x_\ell$ occurring in a clause φ_j, $1 \leq j \leq k$, $\mathcal{P}_\Phi$ contains the atom $\mathsf{D}(x_\ell, j, i)$ with probability 1, where $i = 1$, if the literal is positive, and $i = 0$, if the literal is negative.
- $\mathcal{P}_\Phi$ contains all the atoms $\mathsf{T}(0), \mathsf{S}(0, 1), \mathsf{S}(1, 2), \ldots, \mathsf{S}(k-1, k), \mathsf{K}(k)$, each with probability 1.

We now describe the program Σ_Φ. To detect when a clause is satisfied, we use the additional unary predicate E and the TGD

$$\mathsf{L}(x, i), \mathsf{D}(y, j, i) \rightarrow \mathsf{E}(j),$$

which is a universally quantified formula over the variables x, y, i, and j. We still need to ensure that in each world, exactly one of $\mathsf{L}(x, 0)$ and $\mathsf{L}(x, 1)$ holds. The clauses $x_j \vee \neg x_j$ take care of the lower bound; for the variables $x_1, \ldots, x_n$, we use the TGDs

$$\mathsf{L}(x, 0), \mathsf{L}(x, 1) \rightarrow \mathsf{B} \quad \text{and} \quad \mathsf{B}, \mathsf{D}(y, j, i) \rightarrow \mathsf{E}(j).$$

These TGDs ensure that any inconsistent assignment for $x_1, \ldots, x_n$, i.e., one where some x_j is both *true* and *false*, is automatically marked as satisfying the formula, even if the clause $x_j \vee \neg x_j$ is actually not satisfied. Since there are exactly $4^n - 3^n$ such assignments (where both $\mathsf{L}(x_j, 0)$ and $\mathsf{L}(x_j, 1)$ hold for at least one x_j), we can add this number to the probability threshold that we will use in the end. Note that the probability of each individual assignment is 0.25^n, since there are $2n$ relevant L-atoms (the other atoms are fixed to 0 or 1 and do not contribute here).

It remains to detect whether *all* clauses of φ are satisfied by a consistent assignment, which we do by the means of the TGDs

$$\mathsf{T}(i), \mathsf{S}(i, j), \mathsf{E}(j) \rightarrow \mathsf{T}(j) \quad \text{and} \quad \mathsf{T}(i), \mathsf{K}(i) \rightarrow \mathsf{Z}(i).$$

Lastly, we define the simple UCQ $Q := \exists i \, \mathsf{Z}(i)$. Then, we prove the following claim.

Claim. $\mathsf{P}_{\mathcal{P}_\Phi}(Q, \Sigma_\Phi) \geq 0.25^n(4^n - 3^n + c)$ holds if and only if Φ is valid.
Suppose that Φ is valid, i.e., there are at least c different assignments to $x_1, \ldots, x_n$ that satisfy φ. Then, for each such assignment τ, we can define a world $\mathcal{D}_\tau$ such that it contains all atoms from $\mathcal{P}_\Phi$ that occur with probability 1. Moreover, $\mathcal{D}_\tau$ contains an atom $\mathsf{L}(x_j, 1)$, if x_j is mapped to true in μ, and an atom $\mathsf{L}(x_j, 0)$, if x_j is mapped to false in μ. It is easy to see that each such database $\mathcal{D}_\tau$ is induced by the PDB $\mathcal{P}_\Phi$ and that $\mathcal{D}_\tau \models Q$ by our constructions. In particular, this implies that $\mathcal{D}_\tau \models Q_\Phi$ for c worlds. Recall also that $(4^n - 3^n)$ worlds, capturing the inconsistent valuations, satisfy the query. As every world has the probability $(0.5)^{2n}$, we conclude that $\mathsf{P}_{\mathcal{P}_\Phi}(Q_\Phi) \geq 0.5^{2n}(4^n - 3^n + c)$.

Table 8. The probabilistic database $\mathcal{P}_v$.

Vegetarian		FriendOf		Eats		Meat	
alice	0.7	alice bob	0.7	bob spinach	0.7	shrimp	0.7
bob	0.9	alice chris	0.8	chris mussels	0.8	mussels	0.9
chris	0.6	bob chris	0.1	alice broccoli	0.2	seahorse	0.3

Conversely, if the query probability exceeds the threshold value, then some worlds in $\mathcal{P}_\Phi$ with non-zero probability entail (Q, Σ_Φ), i.e., all clauses of φ are satisfied. Each of the non-zero worlds in PDBs represents a unique combination of atoms of the form $\mathsf{L}(x,0)$ and $\mathsf{L}(x,1)$. The worlds where for at least one variable x_j, $1 \le j \le n$, neither $\mathsf{L}(x_j,0)$ nor $\mathsf{L}(x_j,1)$ holds do not satisfy φ, and hence do not entail (Q, Σ_Φ) and are not counted. Excluding $(4^n - 3^n)$ worlds capturing the inconsistent valuations, all other worlds represent the actual assignments for $x_1, \ldots, x_n$, and hence we know that at least c of those satisfy φ. Thus, we conclude that Φ is valid.

Observe that all TGDs used in the reduction are *full* and *guarded*. Moreover, only the PDB and the probability threshold depend on the input formula (which is allowed in data complexity). Hence, the reduction shows PP-hardness of $\mathsf{PQE}(\mathsf{GF}, \mathsf{UCQ})$ for PDBs in data complexity. $\qquad\square$

GF is one of the least expressive Datalog$^\pm$ languages with polynomial time data complexity for OMQA. Thus, this result already implies PP-hardness for the classes G, F, WS, and WA. This completes all results regarding the data complexity. It remains open whether the data complexity dichotomy can be extended to Datalog-rewritable languages. Clearly, a data complexity dichotomy in these languages would be closely related to a similar result in Datalog.

4.3 Most Probable Explanations for Ontology-Mediated Queries

Motivated by maximal posterior computations in PGMs, we studied the most probable database and most probable hypothesis problems in PDBs. We now extend these results towards ontology-mediated queries. In a nutshell, we restrict ourselves to unions of conjunctive queries (instead of first-order queries), but in exchange consider additional knowledge encoded through an ontology.

Importantly, in MPD (resp., MPH), we are interested in finding the database (resp., the hypothesis) that maximizes the query probability. In the presence of ontological rules, we need to ensure that the chosen database is at the same time consistent with the ontology. More precisely, for ontology-mediated queries, models must be consistent with the ontology; thus, the definitions of MPD and MPH are adapted accordingly.

The Most Probable Database Problem for Ontological Queries. In the most probable database problem for ontological queries, we consider only the *consistent* worlds induced by the PDB, and thus maximize only over consistent worlds.

Definition 13. Let $\mathcal{P}$ be a probabilistic database and Q a query. The *most probable database* for an OMQ (Q, Σ) over a PDB $\mathcal{P}$ is given by

$$\underset{\mathcal{D} \models (Q, \Sigma), mods(\mathcal{D}, \Sigma) \neq \emptyset}{\arg \max} P(\mathcal{D}),$$

where $\mathcal{D}$ ranges over all worlds induced by $\mathcal{P}$.

To illustrate the semantics, we now revisit the Example 4 and the PDB $\mathcal{P}_v$, which is depicted in Table 8.

Example 12. Recall the following query

$$Q_{\mathsf{veg}} := \forall x, y \ \neg\mathsf{Vegetarian}(x) \vee \neg\mathsf{Eats}(x, y) \vee \neg\mathsf{Meat}(y).$$

The most probable database for Q_{veg} contains all atoms from $\mathcal{P}_v$ that have a probability above 0.5, except for $\mathsf{Vegetarian}(\mathsf{chris})$. We can impose the same constraint through the negative constraint

$$\forall x, y \ \mathsf{Vegetarian}(x) \wedge \mathsf{Eats}(x, y) \wedge \mathsf{Meat}(y) \to \bot,$$

and then the most probable database for the query $\top$ will be the same as before. Obviously, we can additionally impose constraints in the form of TGDs, such as

$$\forall x \ \mathsf{Vegetarian}(x) \to \exists y \ \mathsf{FriendOf}(x, y) \wedge \mathsf{Vegetarian}(y),$$

which states that vegetarians have friends who are themselves vegetarians. ∎

The corresponding decision problem is defined as before with the only difference that now we consider ontology-mediated queries.

Definition 14 (MPD). Let (Q, Σ) be an OMQ, $\mathcal{P}$ a probabilistic database, and $p \in (0, 1]$ a threshold. MPD is the problem of deciding whether there exists a database $\mathcal{D}$ that entails (Q, Σ) with $P(\mathcal{D}) > p$. MPD is parametrized with the language of the ontology and the query; we write $\mathsf{MPD}(\mathsf{Q}, \mathcal{L})$ to define MPD on the class Q of queries and on the class of ontologies restricted to the languages $\mathcal{L}$.

We start our complexity analysis with some simple observations. Note first that consistency of a database $\mathcal{D}$ with respect to a program Σ can be written as $\mathcal{D} \not\models (\bot, \Sigma)$, or equivalently, $\mathcal{D} \not\models (Q_\bot, \Sigma^+)$, where $Q_\bot$ is the UCQ obtained from the disjunction of the bodies of all NCs in Σ, and Σ^+ is the corresponding positive program (that contains all TGDs of Σ). This transformation allows us to rewrite the NCs into the query.

A naïve approach to solve MPD is to first guess a database $\mathcal{D}$, and then check that it entails the given OMQ does *not* entail the query $(Q_\bot, \Sigma^+)$ (i.e., it is consistent with the program), and exceeds the probability threshold. Since the probability can be computed in polynomial time, the problem can be decided by a nondeterministic Turing machine using an oracle to check OMQA. Obviously, MPD is at least as hard as OMQA in the underlying ontology languages. These observations result in the following theorem.

Theorem 12. *Let* $\mathfrak{C}$ *denote the data complexity of ontology-mediated query answering for a Datalog$^{\pm}$ language* $\mathcal{L}$. MPD(UCQ, $\mathcal{L}$) *is* $\mathfrak{C}$-*hard and in* NP$^{\mathfrak{C}}$ *under the same complexity assumptions.*

By a reduction from 3-colorability, it is possible to show that MPD is NP-hard already in data complexity for OMQs, even if we only use NCs, i.e., the query and the positive program Σ^+ are empty. This strengthens the previous result about ∀FO queries, since NCs can be expressed by universal queries, but are not allowed to use negated atoms.

Theorem 13. MPD(UCQ, NC) *is* NP-*complete in data complexity (which holds even for instance queries).*

Proof. The upper bound is easy to obtain; thus, we only show the lower bound. We provide a reduction from the well-known 3-colorability problem: given an undirected graph $G = (V, E)$, decide whether the nodes of G are 3-colorable. We first define the PDB $\mathcal{P}_G$ as follows. For all edges $(u, v) \in E$, we add the atom E(u, v) with probability 1, and for all nodes $u \in V$, we add the atoms V$(u, 1)$, V$(u, 2)$, V$(u, 3)$, each with probability 0.7. In this encoding, the atoms V$(u, 1)$, V$(u, 2)$, V$(u, 3)$ correspond to different colorings of the same node u.

We next define the conditions for 3-colorability through a set Σ containing only negative constraints (that do not depend on G). We need to ensure that each node is assigned *at most* one color, which is achieved by means of the negative constraints:

$$V(x, 1), V(x, 2) \rightarrow \bot, \quad V(x, 1), V(x, 3) \rightarrow \bot, \quad V(x, 2), V(x, 3) \rightarrow \bot.$$

Similarly, we need to enforce that the neighboring nodes are not assigned the same color, which we ensure with the negative constraint:

$$E(x, y), V(x, c), V(y, c) \rightarrow \bot.$$

Finally, we define the query $Q := \top$ and prove the following claim.

Claim. G is 3-colorable if and only if there exists a database $\mathcal{D}$ such that $\mathcal{D} \not\models (\bot, \Sigma)$ and $P(\mathcal{D}) \geq (0.7 \cdot 0.3 \cdot 0.3)^{|V|}$.

Suppose that there exists a database $\mathcal{D}$ with a probability of at least $(0.7 \cdot 0.3 \cdot 0.3)^{|V|}$ that satisfies all NCs in Σ. Then, for every node $u \in V$, $\mathcal{D}$ must contain exactly one tuple V(u, c) for some color $c \in \{1, 2, 3\}$. Recall that *at most* was ensured by the first three NCs; hence, in order to achieve the given threshold, *at least* one of these tuples must be present. This yields a unique coloring for the nodes. Furthermore, since $\mathcal{D}$ must contain all tuples corresponding to the edges of G, and $\mathcal{D}$ satisfies the last NC, we conclude that G is 3-colorable.

Suppose that G is 3-colorable. Then, for a valid coloring, we define a DB $\mathcal{D}$ that contains all tuples that correspond to the edges, and add all tuples V(u, c) where c is the color of u. It is easy to see that $\mathcal{D}$ is consistent with Σ and $P(\mathcal{D}) = (0.7 \cdot 0.3 \cdot 0.3)^{|V|}$, which concludes the proof. □

The Most Probable Hypothesis Problem for Ontological Queries. As before, we have to update the definition of most probable hypothesis to take into account only consistent worlds.

Definition 15. The *most probable hypothesis* for an OMQ (Q, Σ) over a PDB $\mathcal{P}$ is

$$\underset{\mathcal{H} \models (Q, \Sigma)}{\arg\max} \sum_{\substack{\mathcal{D} \supseteq \mathcal{H} \\ mods(\mathcal{D}, \Sigma) \neq \emptyset}} P(\mathcal{D}),$$

where $\mathcal{H}$ is a set of (non-probabilistic) atoms t occurring in $\mathcal{P}$.

The corresponding decision problem is defined as before, but parametrized with ontology-mediated queries.

Definition 16 (MPH.) Let (Q, Σ) be an OMQ, $\mathcal{P}$ a PDB, and $p \in (0, 1]$ a threshold. MPH is the problem of deciding whether there exists a hypothesis $\mathcal{H}$ that satisfies (Q, Σ) with $P(\mathcal{H}) > p$. MPH is parametrized with the language of the ontology and the query; we write $MPH(Q, \mathcal{L})$ to define MPH on the class Q of queries and on the class of ontologies restricted to the languages $\mathcal{L}$.

Importantly, computing the probability of a hypothesis for an OMQ is not as easy as for classical database queries, since now there are also inconsistent worlds that shall be ruled out. As a consequence, computing the probability of a hypothesis becomes PP-hard, which makes a significant difference in the complexity analysis.

To solve MPH for OMQs, one can guess a hypothesis, and then check whether it entails the query, and whether the probability mass of its consistent extensions exceeds the given threshold. The latter part can be done by a PP Turing machine with an oracle for OMQA. The oracle can be used also for the initial entailment check. Clearly, $MPH(UCQ, \mathcal{L})$ is at least as hard as OMQA in $\mathcal{L}$. The next theorem follows from these observations.

Theorem 14. Let $\mathfrak{C}$ denote the data complexity of ontology-mediated query answering for a Datalog$^\pm$ language $\mathcal{L}$. $MPD(UCQ, \mathcal{L})$ is $\mathfrak{C}$-hard and in $NP^{PP^{\mathfrak{C}}}$ under the same complexity assumptions.

For any $\mathfrak{C} \subseteq PH$, this result yields $NP^{PP^{PH}} = NP^{PP^{PP}} = NP^{PP}$ as an upper bound, due to a result from [69]. For FO-rewritable Datalog$^\pm$ languages, we immediately obtain an NP^{PP} upper bound for MPH as a consequence of Theorem 14. The obvious question is whether this is also a matching lower bound for FO-rewritable languages?

In general, the (oracle) test of checking whether the probability of a hypothesis exceeds the threshold value is PP-hard. Thus, PP-hardness for MPH for FO-rewritable languages cannot be avoided in general. The remaining question is whether the hypothesis can be identified efficiently, or do we really need to guess the hypothesis? Fortunately, this can be avoided, since we can walk through polynomially many hypotheses (in data complexity) and combine the threshold tests for each of the hypotheses into a single PP computation using the results of [6].

Theorem 15. *Let $\mathcal{L}$ be a Datalog$^\pm$ languages, for which ontology-mediated query answering is in AC^0 in data complexity. Then, $\mathsf{MPH}(\mathsf{UCQ}, \mathcal{L})$ is PP-complete in data complexity under polynomial time Turing reductions.*

Proof. PP-hardness follows from the complexity of probabilistic query evaluation over PDBs ([68], Corollary 1), since we can choose $Q = \top$ and reformulate any UCQ into a set of NCs such that the consistency of a database is equivalent to the non-satisfaction of the UCQ.

We consider an OMQ (Q, Σ), a PDB $\mathcal{P}$, and a threshold p. Since the query is FO-rewritable, it is equivalent to an ordinary UCQ Q_Σ over $\mathcal{P}$. Similarly, we can rewrite the UCQ $Q_\perp$ expressing the non-satisfaction of the NCs into a UCQ $Q_{\perp, \Sigma}$. By the observation in Theorem 5, we can enumerate all hypotheses $\mathcal{H}$, which are the polynomially many matches for Q_Σ in $\mathcal{P}$, and then have to check for each $\mathcal{H}$ whether the probability of all consistent extensions exceeds p. The latter part is equivalent to evaluating $\neg Q_{\perp, \Sigma} \wedge \bigwedge_{t \in \mathcal{H}} t$ over $\mathcal{P}$, which can be done by a PP oracle. We accept if and only if one of these PP checks yields a positive answer. In the terminology of [6], this is a polynomial-time disjunctive reduction of our problem to a PP problem. Since that paper shows that PP is closed under such reductions, we obtain the desired PP upper bound. $\qquad\square$

Given Theorem 15, one may wonder whether the PP upper bound for MPH also applies to languages where OMQA is P-complete in data complexity. Interestingly, $\mathsf{MPH}(\mathsf{UCQ}, \mathsf{GF})$ is already $\mathrm{NP}^{\mathrm{PP}}$-hard.

Theorem 16. $\mathsf{MPH}(\mathsf{UCQ}, \mathsf{GF})$ *is $\mathrm{NP}^{\mathrm{PP}}$-hard in data complexity.*

Proof. We reduce the following problem from [74]: decide the validity of

$$\Phi = \exists x_1, \ldots, x_n \; \mathsf{C}^c \, y_1, \ldots, y_m \, \varphi,$$

where $\varphi = \varphi_1 \wedge \cdots \wedge \varphi_k$ is a propositional formula in CNF, defined over the variables $x_1, \ldots, x_n, y_1, \ldots, y_m$. This amounts to checking whether there is a partial assignment for $x_1, \ldots, x_n$ that admits at least c extensions to $y_1, \ldots, y_m$ that satisfy φ.

We can assume without loss of generality that each of the clauses φ_i contains exactly three literals: shorter clauses can be padded by copying existing literals, and longer clauses can be abbreviated using auxiliary variables that are included under the counting quantifier C^c. Since the values of these variables are uniquely determined by the original variables, this does not change the number of satisfying assignments.

We define the PDB $\mathcal{P}_\Phi$ that describes the structure of Φ:

- For each variable v occurring in Φ, $\mathcal{P}_\Phi$ contains the tuples $\langle \mathsf{V}(v, 0) : 0.5 \rangle$ and $\langle \mathsf{V}(v, 1) : 0.5 \rangle$, where v is viewed as a constant. These tuples represent the assignments that map v to *false* and *true*, respectively.
- For each clause φ_j, we introduce the tuple $\langle \mathsf{C}(v_1, t_1, v_2, t_2, v_3, t_3) : 1 \rangle$, where t_i is 1, if v_i occurs negatively in φ_j, and 0, otherwise (again, all terms are constants). For example, for $x_3 \vee \neg y_2 \vee x_7$, we use $\langle \mathsf{C}(x_3, 0, y_2, 1, x_7, 0) : 1 \rangle$. This encodes the knowledge about the partial assignments that do not satisfy φ.

– We use auxiliary atoms $\langle A(x_1) : 1\rangle$, $\langle S(x_1, x_2) : 1\rangle$, ..., $\langle S(x_{n-1}, x_n) : 1\rangle$, $\langle L(x_n) : 1\rangle$ to encode the order on the variables x_i, and similarly for y_j we use the atoms $\langle B(y_1) : 1\rangle$, $\langle S(y_1, y_2) : 1\rangle$, ..., $\langle S(y_{m-1}, y_m) : 1\rangle$, and $\langle L(y_m) : 1\rangle$.

We now describe the program Σ used for the reduction. First, we detect whether all variables x_i ($1 \leq i \leq n$) have a truth assignment (i.e., at least one of the facts $V(x_i, 0)$ or $V(x_i, 1)$ is present) by the special nullary predicate A, using the auxiliary unary predicates V and A:

$$V(x, t) \rightarrow V(x),$$
$$A(x) \wedge V(x) \wedge S(x, x') \rightarrow A(x'),$$
$$A(x) \wedge V(x) \wedge L(x) \rightarrow A,$$

where x, x', t are variables. We do the same for the variables $y_1, \ldots, y_m$:

$$B(y) \wedge V(y) \wedge S(y, y') \rightarrow B(y'),$$
$$B(y) \wedge V(y) \wedge L(y) \rightarrow B.$$

Now, the query $Q = A$ ensures that only such hypotheses are valid that at least contain a truth assignment for the variables $x_1, \ldots, x_n$.

Next, we restrict the assignments to satisfy φ by using additional NCs in Σ. First, we ensure that there is no "inconsistent" assignment for any variable v, i.e., only one of the facts $V(v, 0)$ or $V(v, 1)$ holds:

$$V(v, 0) \wedge V(v, 1) \rightarrow \bot.$$

Furthermore, if all variables $y_1, \ldots, y_m$ have an assignment, then none of the clauses in φ can be falsified:

$$C(v_1, t_1, v_2, t_2, v_3, t_3) \wedge V(v_1, t_1) \wedge V(v_2, t_2) \wedge V(v_3, t_3) \wedge B \rightarrow \bot,$$

where $v_1, t_1, v_2, t_2, v_3, t_3$ are variables. We now prove the following claim.

Claim. Φ is valid if and only if there exists a hypothesis $\mathcal{H}$ that satisfies (Q, Σ) such that all consistent databases that extend $\mathcal{H}$ sum up to a probability (under $\mathcal{P}_\Phi$) of at least $p = 0.25^n \cdot 0.25^m (3^m - 2^m + c)$.

Assume that such a hypothesis $\mathcal{H}$ exists. Since $\mathcal{H} \models (Q, \Sigma)$, we know that for each x_i ($1 \leq i \leq n$) one of the atoms $V(x_i, 0)$, $V(x_i, 1)$ is included in $\mathcal{H}$. In each consistent extension of $\mathcal{H}$, it must be the case that the complementary facts (representing an inconsistent assignment for x_i) are false. In particular, these complementary facts cannot be part of $\mathcal{H}$, since then its probability would be 0. Hence, we can ignore the factor 0.25^n in the following. There are exactly $3^m - 2^m$ databases satisfying $\mathcal{H}$ that represent consistent, but incomplete assignments for the variables y_j. Since these databases do not entail B, they are all consistent, and hence counted towards the total sum. The inconsistent assignments for $y_1, \ldots, y_m$ yield inconsistent databases, which leaves us only with the 2^m databases representing proper truth assignments. Those that violate at least one clause of φ become inconsistent, and hence there are at least c such consistent databases if and only if there are at least c extensions of the assignment represented by $\mathcal{H}$ that satisfy φ. We conclude these arguments by noting that the probability of each individual choice of atoms $\mathsf{V}(y_j, t_j)$ $(1 \leq j \leq m)$ is 0.25^m.

On the other hand, if Φ is valid, then we can use the same arguments to construct a hypothesis $\mathcal{H}$ (representing the assignment for $x_1, \ldots, x_n$) that exceeds the given threshold. $\square$

We observe a sharp contrast in data complexity results for MPH: PP vs. NP^{PP}, as summarized in Theorems 15 and 16, respectively. These results entail an interesting connection to the data complexity dichotomy for probabilistic query evaluation in PDBs [21]. Building on Theorem 15, we can show a direct reduction from MPH for FO-rewritable languages to probabilistic query evaluation in PDBs, which implies a data complexity dichotomy between P and PP [21].

Theorem 17 (dichotomy). *For FO-rewritable languages $\mathcal{L}$, MPH(UCQ, $\mathcal{L}$) is either in P or PP-hard in data complexity under polynomial time Turing reductions.*

Proof. Recall the proof of Theorem 15. According to [21], the evaluation problem for the UCQ $Q_{\perp, \Sigma}$ over a PDB $\mathcal{P}$ is either in P or PP-hard (under polynomial time Turing reductions). In the former case, MPH can also be decided in deterministic polynomial time. In the latter case, we reduce the evaluation problem for $Q_{\perp, \Sigma}$ over a PDB $\mathcal{P}$ to the MPH for Q_Σ and $Q_{\perp, \Sigma}$ over some PDB $\hat{\mathcal{P}} \supseteq \mathcal{P}$.

For the reduction, we introduce an "artificial match" for Q_Σ into $\hat{\mathcal{P}}$, by adding new constants and atoms (with probability 1) that satisfy one disjunct of Q_Σ, while taking care that these new atoms do not satisfy $Q_{\perp, \Sigma}$. Such atoms must exist if Q_Σ is not subsumed by $Q_{\perp, \Sigma}$; otherwise, all hypotheses would trivially have the probability 0 (and hence the MPH would be decidable in polynomial time). In $\hat{\mathcal{P}}$, the probability of the most probable hypothesis for Q_Σ and $Q_{\perp, \Sigma}$ is the same as the probability of $Q_{\perp, \Sigma}$ over $\mathcal{P}$, and hence deciding the threshold is PP-hard by [21] and Corollary 1. $\square$

5   Related Work

Research on probabilistic databases is almost as old as traditional databases as stated in [68]. We note the seminal work of [31], which has been very important in probabilistic database research as well as the first formulation of possible worlds semantics in the context of databases by [41]. Note that the possible worlds semantics has been widely employed in artificial intelligence; e.g., in probabilistic graphical models [23,44,54] and probabilistic logic programming [24,57,63,64].

Recent advances on PDBs are mainly driven by the dichotomy result given for unions of conjunctive queries [21]. Recently, *open-world probabilistic databases* have been proposed as an alternative open-world probabilistic data model, and the data complexity dichotomy has been lifted to this open-world semantics

[15,16]. Other dichotomy results extend the dichotomy for unions of conjuncive queries in other directions; some allow for disequality ($\neq$) joins in the queries [51] and some for inequality ($<$) joins in the queries [52]. There is also a trichotomy result over queries with aggregation [59]. The common ground in all these dichotomy results is the fact that they classify queries as being safe or unsafe (while the data is not fixed). A different approach is to obtain a classification relative to the structure of the underlying database, and it has been proven in [2], for instance, that every query formulated in monadic second-order logic can be evaluated in linear time over PDBs with a bounded tree-width.

The literature on probabilistic extensions of ontology languages is rich, and we refer the interested reader to a survey [49], and focus on the most recent, and in particular, data-oriented models, which are also recently surveyed in terms of semantic expressivity [9]. Ontology-mediated queries for probabilistic databases have been investigated in the context of both description logics [18,43] and Datalog$^\pm$ [8], and the dichotomy results from PDBs are lifted. Most of the recent work on probabilistic query answering using ontologies is based on lightweight ontology languages, such as the approaches to *Bayesian description logics* in [18,19,22], which combine the description logics of the *DL-Lite* family and the description logic $\mathcal{EL}$, respectively, with Bayesian networks [54]. The underlying probabilistic semantics can be generalized to other ontology languages and PGMs as well. For example, a closely related approach is the one to probabilistic Datalog$^\pm$ in [33], which combines Datalog$^\pm$ with Markov logic networks [61]. In [17], the computational complexity of query answering in probabilistic Datalog$^\pm$ under the possible worlds semantics is investigated.

Maximal posterior computational problems are inspired by PGMs [44], in particular, Bayesian networks [54]. The most probable database problem is introduced in [37], while the most probable hypothesis problem is introduced in [14]. Moreover, these problems are also studied for ontological queries in [14].

6   Conclusion

We have surveyed several query answering and reasoning tasks that can be used to exploit the full potential of probabilistic knowledge bases. In the first part of the tutorial, we focused on (tuple-independent) probabilistic databases as the simplest probabilistic data model. In the second part of the tutorial, we moved on to richer representations where the probabilistic database is extended with ontological knowledge. For each part, we surveyed some known data complexity results and highlighted some recent results.

Acknowledgments. This work was supported by The Alan Turing Institute under the UK EPSRC grant EP/N510129/1, and by the EPSRC grants EP/R013667/1, EP/L012138/1, and EP/M025268/1.

References

1. Abiteboul, S., Hull, R., Vianu, V. (eds.): Foundations of Databases: The Logical Level, 1st edn. Addison-Wesley Longman Publishing Co., Inc., Boston (1995)

2. Amarilli, A., Bourhis, P., Senellart, P.: Tractable lineages on treelike instances: limits and extensions. In: Proceedings of the 35th ACM SIGMOD-SIGACT-SIGAI Symposium on Principles of Database Systems (PODS-16), pp. 355–370. ACM (2016)

3. Baader, F., Calvanese, D., McGuinness, D.L., Nardi, D., Patel-Schneider, P.F. (eds.): The Description Logic Handbook: Theory, Implementation, and Applications, 2nd edn. Cambridge University Press, Cambridge (2007)

4. Baget, J.F., Mugnier, M.L., Rudolph, S., Thomazo, M.: Walking the complexity lines for generalized guarded existential rules. In: Proceedings of the 22nd International Joint Conference on Artificial Intelligence (IJCAI 2011), pp. 712–717 (2011)

5. Beeri, C., Vardi, M.Y.: The implication problem for data dependencies. In: Even, S., Kariv, O. (eds.) ICALP 1981. LNCS, vol. 115, pp. 73–85. Springer, Heidelberg (1981). https://doi.org/10.1007/3-540-10843-2_7

6. Beigel, R., Reingold, N., Spielman, D.: PP is closed under intersection. J. Comput. Syst. Sci. **50**(2), 191–202 (1995)

7. Bienvenu, M., Cate, B.T., Lutz, C., Wolter, F.: Ontology-based data access: A study through disjunctive Datalog, CSP, and MMSNP. ACM Trans. Database Syst. (TODS) **39**(4), 33:1–33:44 (2014)

8. Borgwardt, S., Ceylan, İ.İ., Lukasiewicz, T.: Ontology-mediated queries for probabilistic databases. In: Proceedings of the 31th AAAI Conference on Artificial Intelligence (AAAI 2017), pp. 1063–1069. AAAI Press (2017)

9. Borgwardt, S., Ceylan, İ.İ., Lukasiewicz, T.: Recent advances in querying probabilistic knowledge bases. In: Lang, J. (ed.) Proceedings of the 27th International Joint Conference on Artificial Intelligence and the 23rd European Conference on Artificial Intelligence, IJCAI-ECAI 2018. IJCAI/AAAI Press (2018)

10. Calì, A., Gottlob, G., Kifer, M.: Taming the infinite chase: Query answering under expressive relational constraints. J. Artif. Intell. Res. **48**, 115–174 (2013)

11. Calì, A., Gottlob, G., Lukasiewicz, T.: A general Datalog-based framework for tractable query answering over ontologies. J. Web Semant. **14**, 57–83 (2012)

12. Calì, A., Gottlob, G., Pieris, A.: Towards more expressive ontology languages: The query answering problem. Artif. Intell. **193**, 87–128 (2012)

13. Ceylan, İ.İ.: Query answering in probabilistic data and knowledge bases. Ph.D. thesis, Technische Universität Dresden (2017)

14. Ceylan, İ.İ., Borgwardt, S., Lukasiewicz, T.: Most probable explanations for probabilistic database queries. In: Proceedings of the 26th International Joint Conference on Artificial Intelligence (IJCAI 2017), pp. 950–956. AAAI Press (2017)

15. Ceylan, İ.İ., Darwiche, A., Van den Broeck, G.: Open-world probabilistic databases. In: Proceedings of the 15th International Conference on Principles of Knowledge Representation and Reasoning (KR 2016), pp. 339–348. AAAI Press (2016)

16. Ceylan, İ.İ., Darwiche, A., Van den Broeck, G.: Open-world probabilistic databases: an abridged report. In: Proceedings of the 26th International Joint Conference on Artificial Intelligence (IJCAI 2017), pp. 4796–4800. AAAI Press (2017)

17. Ceylan, İ.İ., Lukasiewicz, T., Peñaloza, R.: Complexity results for probabilistic Datalog±. In: Proceedings of the 28th European Conference on Artificial Intelligence (ECAI 2016). Frontiers in Artificial Intelligence and Applications, vol. 285, pp. 1414–1422. IOS Press (2016)

18. Ceylan, İ.İ., Peñaloza, R.: Probabilistic query answering in the Bayesian description logic $\mathcal{BEL}$. In: Beierle, C., Dekhtyar, A. (eds.) SUM 2015. LNCS (LNAI), vol. 9310, pp. 21–35. Springer, Cham (2015). https://doi.org/10.1007/978-3-319-23540-0_2

19. Ceylan, İ.İ., Peñaloza, R.: The Bayesian ontology language $\mathcal{BEL}$. J. Autom. Reason. **58**(1), 67–95 (2017)

20. Dalvi, N., Suciu, D.: Efficient query evaluation on probabilistic databases. VLDB J. **16**(4), 523–544 (2007)
21. Dalvi, N., Suciu, D.: The dichotomy of probabilistic inference for unions of conjunctive queries. J. ACM **59**(6), 1–87 (2012)
22. d'Amato, C., Fanizzi, N., Lukasiewicz, T.: Tractable reasoning with Bayesian description logics. In: Greco, S., Lukasiewicz, T. (eds.) SUM 2008. LNCS (LNAI), vol. 5291, pp. 146–159. Springer, Heidelberg (2008). https://doi.org/10.1007/978-3-540-87993-0_13
23. Darwiche, A.: Modeling and Reasoning with Bayesian Networks. Cambridge University Press, Cambridge (2009)
24. De Raedt, L., Kimmig, A., Toivonen, H.: ProbLog: A probabilistic prolog and its application in link discovery. In: Proceedings of the 20th International Joint Conference on Artificial Intelligence (IJCAI 2007), pp. 2468–2473. Morgan Kaufmann (2007)
25. Dong, X., Gabrilovich, E., Heitz, G., Horn, W., Lao, N., Murphy, K., Strohmann, T., Sun, S., Zhang, W.: Knowledge Vault: A Web-scale approach to probabilistic knowledge fusion. In: Proceedings of the 20th ACM SIGKDD International Conference on Knowledge Discovery and Data Mining, pp. 601–610. ACM (2014)
26. Dong, X., Gabrilovich, E., Murphy, K., Dang, V., Horn, W., Lugaresi, C., Sun, S., Zhang, W.: Knowledge-based trust: Estimating the trustworthiness of Web sources. Proc. VLDB Endowment **8**(9), 938–949 (2015)
27. Fader, A., Soderland, S., Etzioni, O.: Identifying relations for open information extraction. In: Proceedings of the Conference on Empirical Methods in Natural Language Processing, pp. 1535–1545. Association for Computational Linguistics (2011)
28. Fagin, R., Kolaitis, P.G., Miller, R.J., Popa, L.: Data exchange: Semantics and query answering. Theor. Comput. Sci. **336**(1), 89–124 (2005)
29. Ferrucci, D.A.: Introduction to "This is Watson". IBM J. Res. Dev. **56**(3), 235–249 (2012)
30. Ferrucci, D., Levas, A., Bagchi, S., Gondek, D., Mueller, E.T.: Watson: Beyond Jeopardy!. Artif. Intell. **199–200**, 93–105 (2013)
31. Fuhr, N., Rölleke, T.: A probabilistic relational algebra for the integration of information retrieval and database systems. ACM Trans. Database Syst. (TOIS) **15**(1), 32–66 (1997)
32. Gill, J.T.: Computatonal complexity of probabilistic turing machines. SIAM J. Comput. **6**(4), 675–695 (1977)
33. Gottlob, G., Lukasiewicz, T., Martinez, M.V., Simari, G.I.: Query answering under probabilistic uncertainty in Datalog± ontologies. Ann. Math. Artif. Intell. **69**(1), 37–72 (2013)
34. Grädel, E., Gurevich, Y., Hirsch, C.: The complexity of query reliability. In: Proceedings of the 17th ACM SIGMOD-SIGACT-SIGAI Symposium on Principles of Database Systems (PODS 1998), pp. 227–234. ACM (1998)
35. Greenemeier, L.: Human traffickers caught on hidden internet. Sci. Am. **8** (2015)
36. Gribkoff, E., Suciu, D.: SlimShot: In-database probabilistic inference for knowledge bases. Proc. VLDB Endowment **9**(7), 552–563 (2016)
37. Gribkoff, E., Van den Broeck, G., Suciu, D.: The most probable database problem. In: Proceedings of the 1st International Workshop on Big Uncertain Data (BUDA) (2014)
38. Gribkoff, E., Van den Broeck, G., Suciu, D.: Understanding the complexity of lifted inference and asymmetric weighted model counting. In: Proceedings of the

30th Annual Conference on Uncertainty in Artificial Intelligence (UAI 2014), pp. 280–289. AUAI Press (2014)

39. Hesse, W., Allender, E., Barrington, D.A.M.: Uniform constant-depth threshold circuits for division and iterated multiplication. J. Comput. Syst. Sci. **65**(4), 695–716 (2002)

40. Hoffart, J., Suchanek, F.M., Berberich, K., Weikum, G.: YAGO2: A spatially and temporally enhanced knowledge base from Wikipedia. Artif. Intell. **194**, 28–61 (2013)

41. Imieliski, T., Lipski, W.: Incomplete information in relational databases. J. ACM **31**(4), 761–791 (1984)

42. Jaeger, M.: Relational Bayesian networks. In: Proceedings of the 23rd Annual Conference on Uncertainty in Artificial Intelligence (UAI 1997), pp. 266–273. Morgan Kaufmann (1997)

43. Jung, J.C., Lutz, C.: Ontology-based access to probabilistic data with OWL QL. In: Cudré-Mauroux, P., et al. (eds.) ISWC 2012. LNCS, vol. 7649, pp. 182–197. Springer, Heidelberg (2012). https://doi.org/10.1007/978-3-642-35176-1_12

44. Koller, D., Friedman, N.: Probabilistic Graphical Models: Principles and Techniques. MIT Press, Cambridge (2009)

45. Krötzsch, M., Rudolph, S.: Extending decidable existential rules by joining acyclicity and guardedness. In: Proceedings of the 22nd International Joint Conference on Artificial Intelligence (IJCAI 2011), pp. 963–968. AAAI Press (2011)

46. Ku, J.P., Hicks, J.L., Hastie, T., Leskovec, J., Ré, C., Delp, S.L.: The mobilize center: an NIH big data to knowledge center to advance human movement research and improve mobility. J. Am. Med. Inform. Assoc. **22**(6), 1120–1125 (2015)

47. Libkin, L.: Elements of Finite Model Theory. Springer, Heidelberg (2004). https://doi.org/10.1007/978-3-662-07003-1

48. Littman, M.L., Goldsmith, J., Mundhenk, M.: The computational complexity of probabilistic planning. J. Artif. Intell. Res. **9**, 1–36 (1998)

49. Lukasiewicz, T., Straccia, U.: Managing uncertainty and vagueness in description logics for the Semantic Web. J. Web Semant. **6**(4), 291–308 (2008)

50. Mitchell, T., et al.: Never-ending learning. In: Proceedings of the 29th AAAI Conference on Artificial Intelligence (AAAI 2015), pp. 2302–2310 (2015)

51. Olteanu, D., Huang, J.: Using OBDDs for efficient query evaluation on probabilistic databases. In: Greco, S., Lukasiewicz, T. (eds.) SUM 2008. LNCS (LNAI), vol. 5291, pp. 326–340. Springer, Heidelberg (2008). https://doi.org/10.1007/978-3-540-87993-0_26

52. Olteanu, D., Huang, J.: Secondary-storage confidence computation for conjunctive queries with inequalities. In: Proceedings of the 2009 ACM SIGMOD International Conference on Management of Data, pp. 389–402. ACM (2009)

53. Papadimitriou, C.H.: Computational Complexity. Addison-Wesley, Boston (1994)

54. Pearl, J.: Probabilistic Reasoning in Intelligent Systems. Morgan Kaufmann, Burlington (1988)

55. Peters, S.E., Zhang, C., Livny, M., Ré, C.: A machine reading system for assembling synthetic paleontological databases. PLoS One **9**(12), e113523 (2014)

56. Poggi, A., Lembo, D., Calvanese, D., De Giacomo, G., Lenzerini, M., Rosati, R.: Linking data to ontologies. J. Data Semant. **10**, 133–173 (2008)

57. Poole, D.: The independent choice logic for modelling multiple agents under uncertainty. Artif. Intell. **94**(1–2), 7–56 (1997)

58. Provan, J.S., Ball, M.O.: The complexity of counting cuts and of computing the probability that a graph is connected. SIAM J. Comput. **12**(4), 777–788 (1983)

59. Ré, C., Suciu, D.: The trichotomy of having queries on a probabilistic database. VLDB J. **18**(5), 1091–1116 (2009)
60. Reiter, R.: On closed world data bases. In: Gallaire, H., Minker, J. (eds.) Logic and Data Bases, pp. 55–76. Springer, Heidelberg (1978). https://doi.org/10.1007/978-1-4684-3384-5_3
61. Richardson, M., Domingos, P.: Markov logic networks. Mach. Learn. **62**(1), 107–136 (2006)
62. Rossman, B.: Homomorphism preservation theorems. J. ACM **55**(3), 1–53 (2008)
63. Sato, T.: A statistical learning method for logic programs with distribution semantics. In: Proceedings of the 12th International Conference on Logic Programming (ICLP 1995), pp. 715–729. MIT Press (1995)
64. Sato, T., Kameya, Y.: PRISM: A language for symbolic-statistical modeling. In: Proceedings of the 15th International Joint Conference on Artificial Intelligence (IJCAI 1997), pp. 1330–1335. Morgan Kaufmann (1997)
65. Shin, J., Wu, S., Wang, F., De Sa, C., Zhang, C., Ré, C.: Incremental knowledge base construction using DeepDive. Proc. VLDB Endowment **8**(11), 1310–1321 (2015)
66. Sipser, M.: Introduction to the Theory of Computation, 1st edn. International Thomson Publishing, Boston (1996)
67. Staworko, S., Chomicki, J.: Consistent query answers in the presence of universal constraints. Inf. Syst. **35**(1), 1–22 (2010)
68. Suciu, D., Olteanu, D., Ré, C., Koch, C.: Probabilistic Databases. Synthesis Lectures on Data Management. Morgan & Claypool Publishers, San Rafael (2011)
69. Toda, S.: On the computational power of PP and ⊕P. In: Proceedings of the 30th Annual Symposium on Foundations of Computer Science, pp. 514–519 (1989)
70. Toda, S., Watanabe, O.: Polynomial-time 1-turing reductions from #PH to #P. Theor. Comput. Sci. **100**(1), 205–221 (1992)
71. Valiant, L.G.: The complexity of computing the permanent. Theor. Comput. Sci. **8**(2), 189–201 (1979)
72. Vardi, M.Y.: The complexity of relational query languages. In: Lewis, H.R., Simons, B.B., Burkhard, W.A., Landweber, L.H. (eds.) Proceedings of the 14th Annual ACM Symposium on Theory of Computing (STOC 1982), pp. 137–146. ACM (1982)
73. Vollmer, H.: Introduction to Circuit Complexity: A Uniform Approach. Springer, Heidelberg (1999). https://doi.org/10.1007/978-3-662-03927-4
74. Wagner, K.W.: The complexity of combinatorial problems with succinct input representation. Acta Informatica **23**(3), 325–356 (1986)
75. Wu, W., Li, H., Wang, H., Zhu, K.Q.: Probase: A probabilistic taxonomy for text understanding. In: Proceedings of the 2012 ACM SIGMOD International Conference on Management of Data, pp. 481–492. ACM (2012)

Cold-Start Knowledge Base Population Using Ontology-Based Information Extraction with Conditional Random Fields

Hendrik ter Horst, Matthias Hartung, and Philipp Cimiano[✉]

CITEC, Bielefeld University, Bielefeld, Germany
{hterhors,mhartung,cimiano}@techfak.uni-bielefeld.de

Abstract. In this tutorial we discuss how Conditional Random Fields can be applied to knowledge base population tasks. We are in particular interested in the cold-start setting which assumes as given an ontology that models classes and properties relevant for the domain of interest, and an empty knowledge base that needs to be populated from unstructured text. More specifically, cold-start knowledge base population consists in predicting semantic structures from an input document that instantiate classes and properties as defined in the ontology. Considering knowledge base population as structure prediction, we frame the task as a statistical inference problem which aims at predicting the most likely assignment to a set of ontologically grounded output variables given an input document. In order to model the conditional distribution of these output variables given the input variables derived from the text, we follow the approach adopted in Conditional Random Fields. We decompose the cold-start knowledge base population task into the specific problems of entity recognition, entity linking and slot-filling, and show how they can be modeled using Conditional Random Fields.

Keywords: Cold-start knowledge base population
Ontology-based information extraction · Slot filling
Conditional random fields

1 Introduction

In the era of data analytics, knowledge bases are vital sources for various downstream analytics tasks. However, their manual population may be extremely time-consuming and costly. Given that in many scientific and technical domains, it is still common practice to rely on natural language as the primary medium for knowledge communication, information extraction techniques from natural language processing [17,26] pose a viable alternative towards (semi-)automated knowledge base population by transforming unstructured textual information into structured knowledge.

© Springer Nature Switzerland AG 2018
C. d'Amato and M. Theobald (Eds.): Reasoning Web 2018, LNCS 11078, pp. 78–109, 2018.
https://doi.org/10.1007/978-3-030-00338-8_4

Against this backdrop, *cold-start knowledge base population* [14] has recently attracted increasing attention. Cold-start knowledge base population can be seen as a particular instance of an information extraction problem with two characteristics: First, information extraction serves as an upstream process in order to populate an *initally empty* knowledge base. Second, an ontology is given that defines the structure of a domain of interest in terms of classes and properties (entities and relations). Based on these requirements, the goal is to populate a knowledge base that structurally follows the specifications of the ontology, given a collection of textual data. This implies extracting classes (entities) and filling their properties (as defined by the underlying ontology).

Knowledge base population can be modeled as a statistical inference problem. Given a document as input, the goal is to infer the most likely instantiation(s) of ontological structures that best capture the knowledge expressed in the document. Modeling the cold-start population task as statistical inference problem requires the computation of the distribution of possible outputs. Here, an output refers to a specific variable assignment that determines the instantiation of such structure(s) of interest. In the context of stochastic models, we are in particular interested in the conditional probability of the variables of the output given an input document. Let $\boldsymbol{y} = (y_1, \ldots, y_m)$ specify the output vector of variables and $\boldsymbol{x} = (x_1, \ldots, x_n)$ the input vector of variables (usually tokens of a document). We are interested in modeling the following probability distribution:

$$p(\boldsymbol{y}|\boldsymbol{x}) = p(y_1, \ldots, y_m | x_1, \ldots, x_n)$$

Given a model of this distribution, the goal is to find the assignment that maximizes the likelihood under the model, that is:

$$\hat{y_1}, \ldots, \hat{y_m} = \underset{y_1, \ldots, y_n}{\operatorname{argmax}} \, p(y_1, \ldots, y_m | x_1, \ldots, x_n)$$

Typically, probabilistic models are parameterized by some parameter vector θ that is learned during a training phase:

$$p(y_1, \ldots, y_m | x_1, \ldots, x_n; \theta)$$

One class of machine learning models that provides an efficient computation of the above distribution are called Conditional Random Fields (CRFs; [11,23]). A CRF typically models the probability of hidden output variables conditioned on given observed input variables in a factorized form, that is relying on a decomposition of the probability into local factors. These factors reflect the compatibility of variable assignments in predefined subsets of random variables. Conditional random fields are typically trained in a discriminative fashion with the objective to maximize the likelihood of the data given the parametrized model.

In this tutorial, we discuss how conditional random fields can be applied to two constitutive subtasks of knowledge base population.

Entity Recognition and Linking. As a first task, we show how the problem of entity recognition and linking [6,21] can be modeled. In particular, we investigate the problem of disease recognition and linking from biomedical texts as illustrated in the following example taken from PubMed[1]. We underline occurrences of diseases (recognition), concepts (as defined in the MeSH[2] thesaurus) are shown in subscript (linking):

Example 1. "An instance of <u>aortic intimal sarcoma</u>_{D001157} [...], with clinical evidence of acutely <u>occurring hypertension</u>_{D006973} [...], and <u>aortic occlusion</u>_{D001157} in a 50-year-old male is reported."

The conditional probability of the example can be explicitly expressed as[3] :

$$p(\boldsymbol{y}|\boldsymbol{x}) = p(y_1 = \langle \text{"aortic intimal sarcoma"}, D001157 \rangle,$$
$$y_2 = \langle \text{"occurring hypertension"}, D006973 \rangle,$$
$$y_3 = \langle \text{"aortic occlusion"}, D001157 \rangle \mid$$
$$x_1 = \text{"An"}, x_2 = \text{"instance"}, \ldots,$$
$$x_{n-1} = \text{"reported"}, x_n = \text{"."})$$

Slot Filling. Second, we show how slot filling can be modeled via conditional random fields. We consider slot filling as a relation extraction task with ontologically defined templates as output structures. Such templates consist of a number of typed slots to be filled from unstructured text [3]. Following an ontology-based approach [26], we assume that these templates (including slots and types of their potential fillers) are pre-defined in a given ontology.[4] Consider the following input document:

Example 2. "Six- to eight-week-old adult female (192-268 g) Sprague-Dawley rats were used for these studies."

In this example, we are interested in predicting an *AnimalModel* template as specified by the Spinal Cord Injury Ontology (SCIO) [2]. This ontological template specifies details about the animal model that was used in a pre-clinical study. A probable variable assignment of the output might be:

age → "Six- to eight-week",
age category → Adult,
gender → Female,
weight → "192 - 268 g",
species → Sprague Dawley Rat.

This tutorial paper is structured as follows. Section 2 provides an introduction to conditional random fields as well as inference inference and parameter learning.

---

[1] https://www.ncbi.nlm.nih.gov/pubmed?cmd=search&term=2584179.

[2] https://www.ncbi.nlm.nih.gov/mesh.

[3] Note, the conditional probability can be modeled in many different ways, depending on the model structure.

[4] Considering ontological properties, one must distinguish between object-type and data-type properties. Values for the latter case are arbitrary literals and thus **not** predefined.

In Sect. 3, we apply this approach to the problem of named entity recognition and linking in the biomedical domain, namely for diseases and chemicals. In Sect. 4, we apply our approach to the task of slot filling in the domain of therapies about spinal cord injury and provide all necessary information to tackle this task. This tutorial ends with Sect. 5 in which we conclude our proposed approach. Parts of the materials presented here are taken from our previous publications [4,7,8].

2   Conditional Random Fields for Knowledge Base Population

Many tasks in knowledge base population and natural language processing in general can be modeled as structure prediction problems where ontology-defined target structures need to be predicted from some input structure [22]. A particular case of this are sequence-to-sequence prediction problems such as part-of-speech tagging or named-entity-recognition. Here, an output sequence needs to be predicted from a given sequence of tokens.

From a general perspective, such tasks require predicting a hidden output vector y on the basis of an observed input vector x. Usually x represents a tokenized document (containing natural language) in which all variables $x_t \in x$ correspond to single tokens in the document. Thus, the length of the input vector is equal to the number of tokens T in the document, that is $|x| = T$, where x_t corresponds to the tth token. The hidden output vector y may vary in length and complexity depending on the structure of the problem.

Such problems can be modeled via a conditional distribution of the following form:

$$p(y|x; \theta),$$

where the probability of the output is conditioned on the input and parametrized by some vector θ.

The variable assignment that maximizes the probability can be found by what is called *Maximum A Posteriori* (MAP) inference:

$$\hat{y} = \underset{y}{\operatorname{argmax}}\, p(y|x; \theta).$$

Conditional Random Fields (CRF) are widely applied for such problems as they can model the above conditional distribution via a product of *factors* directly. These factors are parameterized with subsets of $y_i \subseteq y$ and $x_i \subseteq x$. Factors and the corresponding variables are typically specified in a so called *factor graph* [9,10]. A factor graph is a bipartite graph $\mathcal{G} = (V, E, F)$ consisting of a set of random variables V, factors F and edges E. We define $v_j \in V$ as a subset of all possible random variables: $v_j = y_j \cup x_j$. Each factor $\Psi_j \in F$ represents a function: $\Psi_j : V_j \rightarrow \mathbb{R}_{\geq 0}$ that is parameterized with v_j and returns a non-negative scalar score indicating the compatibility of variables in v_j. Further, an edge $e_j \in E$ is defined as a tuple: $e_j = \langle V_j, \Psi_j \rangle$. An important aspect is that CRFs assume x as fully observed and thus do not model statistical dependencies between variables in x. Figure 1 shows an example factor graph with

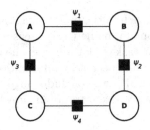

Fig. 1. Bipartite undirected factor graph with $V = \{A, B, C, D\}$ and $F = \{\Psi_1, \Psi_2, \Psi_3, \Psi_4\}$ (black boxes).

$V = \{A, B, C, D\}$ and $F = \{\Psi_1, \Psi_2, \Psi_3, \Psi_4\}$. Based on the structure of this factor graph, the factorization can be formulated as:

$$p(A, B, C, D) = \frac{1}{Z}\Psi_1(A, B) \cdot \Psi_2(A, C) \cdot \Psi_3(C, D) \cdot \Psi_4(B, D) \tag{1}$$

where Z is the partition function that sums up over all possible variable assignments in order to ensure a valid probability distribution:

$$Z = \sum_{a \in A, b \in B, c \in C, d \in D} \Psi_1(a, b) \cdot \Psi_2(a, c) \cdot \Psi_3(c, d) \cdot \Psi_4(b, d). \tag{2}$$

To concretize the example, let each random variable in V can take binary values, that is $A = \{a_1, a_2\}, B = \{b_1, b_2\}, C = \{c_1, c_2\}, D = \{d_1, d_2\}$, and each factor Ψ_i computes a score that reflects the compatibility of two variables as shown in Table 1. The probability for a concrete variable assignment e.g. $A = a_1, B = b_1, C = c_2$ and $D = d_1$ is then explicitly calculated as:

$$p(a_1, b_1, c_2, d_1) = \frac{1}{Z}(\Psi_1(a_1, b_1) \cdot \Psi_2(a_1, c_2) \cdot \Psi_3(c_2, d_1) \cdot \Psi_4(b_1, d_1))$$
$$= \frac{1}{Z}(5 \cdot 4 \cdot 1 \cdot 1) = \frac{1}{Z}20 \tag{3}$$

where

$$Z = \Psi_1(a_1, b_1) \cdot \Psi_2(a_1, c_1) \cdot \Psi_3(c_1, d_1) \cdot \Psi_4(b_1, d_1)$$
$$+ \Psi_1(a_1, b_1) \cdot \Psi_2(a_1, c_1) \cdot \Psi_3(c_1, d_2) \cdot \Psi_4(b_1, d_2)$$
$$+ \dots \tag{4}$$
$$+ \Psi_1(a_2, b_2) \cdot \Psi_2(a_2, c_2) \cdot \Psi_3(c_2, d_2) \cdot \Psi_4(b_2, d_2)$$
$$= 659$$

So that the probability is calculated as: $p(a_1, b_1, c_2, d_1) = \frac{20}{659} = 0.03$.

This is essentially the approach taken by conditional random fields which model the conditional distribution of output variables given input variables through a product of factors that are defined by a corresponding factor graph:

$$p(\mathbf{y}|\mathbf{x}) = \frac{1}{Z(\mathbf{x})} \prod_{\Psi_i \in F} \Psi_i(\mathcal{N}(\Psi_i)), \tag{5}$$

Table 1. Compatibility table for all possible pairwise variable assignments. The specific assignment $A = a_1, B = b_1, C = c_2$, and $D = d_1$ is highlighted.

A B $\Psi_1(\cdot,\cdot)$		A C $\Psi_2(\cdot,\cdot)$		B D $\Psi_3(\cdot,\cdot)$		C D $\Psi_4(\cdot,\cdot)$	
$a_1\ b_1$	5	$a_1\ c_1$	3	$b_1\ d_1$	1	$c_1\ d_1$	3
$a_1\ b_2$	2	$\mathbf{a_1\ c_2}$	4	$b_1\ d_2$	1	$c_1\ d_2$	2
$a_2\ b_1$	2	$a_2\ c_1$	0	$b_2\ d_1$	1	$\mathbf{c_2\ d_1}$	1
$a_2\ b_2$	1	$a_2\ c_2$	3	$b_2\ d_2$	7	$c_2\ d_2$	4

where $\mathcal{N}(\Psi_i)$ is the set of variables neighboring Ψ_i in the factor graph:

$$\mathcal{N}(\Psi_i) := \{v_i \mid (v_i, \Psi_i) \in E\}$$

Typically, factors are log-linear functions specified in terms of feature functions $f_j \in \mathcal{F}_j$ as sufficient statistics:

$$\Psi_i(\mathcal{N}(\Psi_i)) = \exp\left\{\sum_{f_j \in \mathcal{F}_i} f_j(\boldsymbol{y}_i, \boldsymbol{x}_i) \cdot \theta_i\right\}$$

This yields the following general form for a conditional random field that represents the conditional probability distribution:

$$p(\boldsymbol{y}|\boldsymbol{x}) = \frac{1}{Z(\boldsymbol{x})} \prod_{\Psi_i \in F} \exp\left\{\sum_{f_j \in \mathcal{F}_i} f_j(\boldsymbol{y}_i, \boldsymbol{x}_i) \cdot \theta_i\right\} \tag{6}$$

The number of factors is determined by the length of the input and by the output structure as defined by the problem. The number of factors differs in any case by the size of the input. This leads one to consider factor types that are determined by so called *factor templates* (sometimes clique templates) that can be rolled out over the input yielding specific factor instances. Hereby, all the factors instantiating a particular template are assumed to have the same parameter vector θ_Ψ. Each template $C_j \in \mathcal{C}$ defines (i) subsets of observed and hidden variables for which it can generate factors and (ii) feature functions to provide sufficient statistics. All factors generated by a template C_j share the same parameters θ_j. With this definition, we reformulate the conditional probability from Eq. (5) as follows:

$$p(\boldsymbol{y}|\boldsymbol{x}; \theta) = \frac{1}{Z(\boldsymbol{x})} \prod_{C_j \in \mathcal{C}} \prod_{\Psi_i \in C_j} \Psi_i(\boldsymbol{y}_i, \boldsymbol{x}_i, \theta_j) \tag{7}$$

Linear Chain CRF. A Linear Chain CRF is a linear structured instance of a factor graph which is mostly used to model sequence-to-sequence problems. The linear chain CRF factorizes the conditional probability under the following restriction. A hidden variable y_t at position $t \in [0..T]$ depends only on itself, the value of the previous hidden variable y_{t-1} and a subset of observed variables $x_t \subseteq x$. x_t contains all information that is needed to compute the factor $\Psi_t(y_t, y_{t-1}, x_t)$ at position t. For example, factors that are based on the context tokens with distance of δ to each side, the observed vector at position t is $x_t = \{x_{t-\delta}, \ldots, x_t, \ldots, x_{t+\delta}\}$. Each factor $\Psi_t \in F$ computes a log-linear value based on the scalar product of a factor related feature vector $\mathcal{F}_t$, to be determined from the corresponding subset of variables, and a set of related parameters θ_t. Due to the linear nature of the factor graph $\mathcal{G}$, feature functions are formulated in the form of $f_t(y_t, y_{t-1}, x_t)$. The decomposed conditional probability distribution is then defined on the joint probability $p(y_t, y_{t-1}, x_t)$ as formulated in Eq. (8):

$$p(\boldsymbol{y}|\boldsymbol{x}) = \frac{1}{Z(\boldsymbol{x})} \prod_{t=1}^{T} \Psi_t(y_t, y_{t-1}, x_t) \tag{8}$$

where each Ψ_t has the log-linear form:

$$\Psi_t(y_t, y_{t-1}, x_t) = \exp\left\{ \sum_{i=1}^{\mathcal{F}_t} f_i(y_t, y_{t-1}, x_t) \cdot \theta_i \right\} \tag{9}$$

An example of a linear chain CRF for a sequence-to-sequence application with $|\boldsymbol{y}| = |\boldsymbol{x}|$ and $x_t = \{x_t\}$ is shown in Fig. 2. Observed variables $\boldsymbol{x}$ is a sequence of tokens from the sentence: "*Barack Obama is the former president of the USA.*".

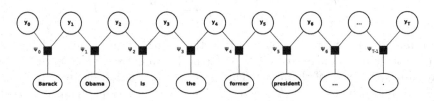

Fig. 2. Linear chain CRF for a sequence-to-sequence application with $|\boldsymbol{y}| = |\boldsymbol{x}|$ and $x_t = \{x_t\}$. Observed variables $\boldsymbol{x}$ are tokens from a tokenized input sentence.

Linear chain CRF models are commonly used for sequence tagging problems where $|\boldsymbol{y}| = |\boldsymbol{x}|$. A well known problem that fits this condition is POS tagging. Given a finite set Φ of possible (POS) tags e.g. $\Phi = \{NNP, VBZ, DT, JJ, NN, IN, .\}$, the goal is to assign a tag to each token in $\boldsymbol{x}$ so that the overall probability of the output tag-sequence $p(\boldsymbol{y})$ is maximized.[5] But also more complex tasks such as named entity recognition can be formulated as a sequence-to-sequence problem although the number of entities

---

[5] Based on the given example, the optimal output vector is $\boldsymbol{y}^* = \{NNP, NNP, VBZ, DT, JJ, NN, IN, DT, NNP, .\}$.

is priorly unknown. For that, the input document is transformed into an IOB-sequence, that is the document is tokenized and each token is labeled with one value of $\Phi = \{I, O, B\}$, (where B states the *beginning* of an entity, I states that the token is *inside* of an entity, and tokens labeled with O are outside of an entity).[6]

2.1 Inference and Learning

Although the factorization of the probability density already reduces the complexity of the model, exact inference is still intractable for probabilistic graphical models in the general case and for conditional random fields in particular. While efficient inference algorithms exist for the case of linear chain CRFs (cf. [23]), in the general case inference requires computing the partition function $Z(x)$ which sums up over an exponential number of possible assignments to the variables $Y_1, ..., Y_n$. Further, inference for a subset $Y_A \subseteq Y$ of variables requires marginalization over the remaining variables in addition to computing the partition function.

Maximum-A-Posteriori Inference (MAP) in turn requires considering all possible assignments to the variables $Y_1, ... Y_n$ to find the maximum.

To avoid the exponential complexity of inference in conditional random fields, often approximative inference algorithms are used. One class of such algorithms are Markov Chain Monte Carlo (MCMC) methods that iteratively generate stochastic samples from a joint distribution $p(y)$ to approximate the posterior distribution.

Samples are probabilistically drawn from a state space Y that contains (all) possible variable assignments (*state*) for y. While walking through the state space, MCMC constructs a Markov Chain that, with sufficient samples, converges against the real distribution of interest. That means the distribution of states within the chain approximates the marginal probability distribution of $p(y_i)$ for all $y_i \in y$. The drawback of this method is that it is priorly unknown how many iterations are needed to ensure convergence.

Inference: In high dimensional multivariate distributions the Markov Chain can be efficiently constructed by Metropolis–Hastings sampling algorithms. In Metropolis–Hastings, new samples are drawn from a probability distribution Q. The next drawn sample y' is only conditioned on the previous sample y making it a Markov Chain. If Q is proportional to the desired distribution p, then, with sufficient samples, the Markov Chain will approximate the desired distribution by using a stochastically-based accept/reject strategy. The pseudo-code for the standard procedure of Metropolis–Hastings is presented in Algorithm 2.1.

Here, the function acceptanceRatio($\cdot, \cdot$) calculates a ratio for a new state to be accepted as the next state. In standard Metropolis–Hastings, this ratio is

---

[6] Based on the given example, the optimal output vector for NER is $y^* = \{B, I, O, O, O, O, O, O, B, O\}$. The generated sequence, tells us that the tokens *Barack* and *Obama* belong to the same entity (B is directly followed by I), whereas *USA* is another single token entity.

Algorithm 1. Pseudo-code Metropolis–Hastings Sampling

1: $y_0 \leftarrow random\ sample$
2: $t \leftarrow 1$
3: **repeat**
4:     $y' \sim \mathcal{Q}(y'|y_t)$
5:     $\alpha \leftarrow \text{acceptanceRatio}(y', y_t)$
6:     **if** $\alpha \geq \text{rand}[0,1]$ **then**
7:         $y_{(t+1)} \leftarrow y'$
8:     **else**
9:         $y_{(t+1)} \leftarrow y$
10:    **end if**
11:    $t \leftarrow t+1$
12: **until** convergence

---

computed as the probability of the new state divided by the probability of the current state:

$$\text{acceptanceRatio}(y', y) = \frac{f(y')}{f(y)}, \tag{10}$$

where $f(y)$ is a function that is proportional the real density $p(y)$. Note that, if $f(y') \geq f(y)$, the new state y' will be always accepted as the resulting ratio is greater 1. Otherwise, the likelihood of being accepted is proportional to the likelihood under the model.

One special case of the general Metropolis–Hastings algorithm is called Gibbs sampling. Instead of computing the fully joint probability of all variables in $p(\boldsymbol{y}) = p(y_1, \ldots, y_n)$ in Gibbs each variable y_i individually resampled while keeping all other variables fixed, that makes $p(y_i|\boldsymbol{y}_{\backslash i})$. Resnik et al. [20] describe, drawing the next Gibbs sample as:

Algorithm 2. Create next sample with Gibbs

1: **for** $i = 1$ to n **do**
2:     $y_i^{(t+1)} \sim p(y_i|y_1^{(t+1)}, \ldots, y_{i-1}^{(t+1)}, y_{i+1}^{(t)}, \ldots, y_n^{(t)})$
3: **end for**

We propose a slightly different sampling procedure (hereinafter called atomic change sampling) as depicted in Fig. 3.

While in standard Gibbs sampling, one needs to specify the order of variables that are resampled, we relax this prerequisite by extending the state space in each intermediate step to all possible states that can be reached by applying one atomic change to the current state. Let $\Omega(\boldsymbol{y})$ be the set of states that can be generated from $\boldsymbol{y}$ by applying one atomic change operation to $\boldsymbol{y}$, then the probability distribution $\mathcal{Q}$ can be described as:

$$\mathcal{Q}(\boldsymbol{y}', \boldsymbol{y}) = \begin{cases} q(\boldsymbol{y}') & \text{iff } \boldsymbol{y}' \in \Omega(\boldsymbol{y}) \\ 0 & \text{else} \end{cases}, \tag{11}$$

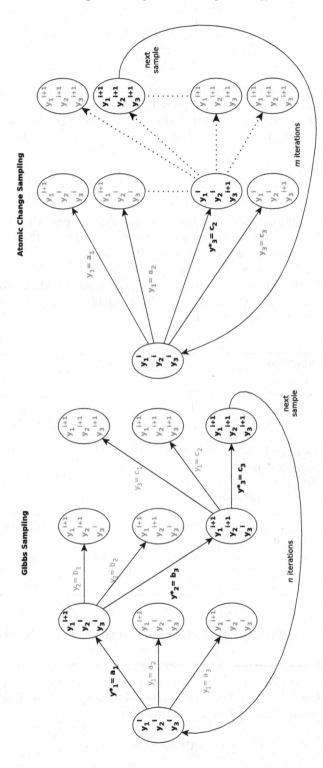

Fig. 3. Comparison of standard Gibbs sampling (left) and atomic change sampling (right). The variable assignment that was drawn and accepted from the distribution of possible assignments is highlighted

where

$$q(\boldsymbol{y}') = \frac{f(\boldsymbol{y}')}{\sum_{\hat{\boldsymbol{y}} \in \Omega(\boldsymbol{y})} f(\hat{\boldsymbol{y}})}. \tag{12}$$

Parameter Learning. The learning problem consists of finding the optimal weight vector θ that maximizes the a-posteriori probability $p(\boldsymbol{y}|\boldsymbol{x};\theta)$.

Typically, the parameters of the distribution are optimized given some training data $D = (\boldsymbol{y}_i, \boldsymbol{x}_i)$ to maximize the likelihood of the data under the model, that is

$$\hat{\theta} = argmax_\theta \prod_{\boldsymbol{y}_i, \boldsymbol{x}_i \in D} P(\boldsymbol{y}_i|\boldsymbol{x}_i, \theta)$$

However, parameter optimization typically calls the inference procedure to estimate the expected count of features under the model θ to compute the gradient that maximizes the likelihood of the data under the model.

Another solution to parameter learning is to rely on a ranking objective that attempts to update the parameter vector to assign a higher likelihood to preferred solutions. This is the approach followed by SampleRank [25]. The implementation in our approach is shown below:

Algorithm 3. Sample Rank

1: **Inputs:** training data D
2: **Initialization:** set $\theta \leftarrow \mathbf{0}$, set $y \leftarrow y_0 \in Y$
3: **Output:** parameter θ
4: **repeat**
5:     $y' \sim \mathbb{M}(\cdot|y)$
6:     $\Delta \leftarrow \phi(y', x) - \phi(y, x)$
7:     **if** $\theta \cdot \Delta > 0 \wedge \mathbb{P}(y, y')$ **then**
8:         $\theta \leftarrow \theta - \eta\Delta$
9:     **else if** $\theta \cdot \Delta \leq 0 \wedge \mathbb{P}(y', y)$ **then**
10:         $\theta \leftarrow \theta + \eta\Delta$
11:     **end if**
12:     **if** accept(y', y) **then**
13:         $y \leftarrow y'$
14:     **end if**
15: **until** convergence

SampleRank is an online algorithm which learns preferences over hypotheses from gradients between atomic changes to overcome the expensive computational costs that arise during inference. The parameter update is based on gradient descent on pairs of states $(\boldsymbol{y}^t, \boldsymbol{y}^{(t+1)})$ consisting of the current best state $\boldsymbol{y}^t$ and the successor state $\boldsymbol{y}^{(t+1)}$. Two states are compared according to the following objective preference function $\mathbb{P} : Y \times Y \rightarrow \{false, true\}$:

$$\mathbb{P}(\boldsymbol{y}, \boldsymbol{y}') = \mathbb{O}(\boldsymbol{y}') > \mathbb{O}(\boldsymbol{y}) \tag{13}$$

Here, $\mathbb{O}(\boldsymbol{y})$ denotes an objective function that returns a score indicating its degree of accordance with the ground truth from the respective training document. $\mathbb{M} : Y \times Y \rightarrow [0, 1]$ denotes the proposal distribution that is provided by the model, $\phi : Y \times X \rightarrow \mathcal{R}^{|\theta|}$ denotes the sufficient statistics of a specific variable assignment and:

$$\text{accept}(y, y') \leftrightarrow p(y') > p(y) \tag{14}$$

3 Conditional Random Fields for Entity Recognition and Linking

As a subtask in machine reading, i.e., automatically transforming unstructured natural language text into structured knowledge [18], entity linking facilitates various applications such as entity-centric search or predictive analytics in knowledge graphs. In these tasks, it is advisable to search for the entities involved at the level of unique knowledge base identifiers rather than surface forms mentioned in the text, as the latter are ubiquitously subject to variation (e.g., spelling variants, semantic paraphrases, or abbreviations). Thus, entities at the concept level can not be reliable, retrieved or extracted from text using exact string match techniques.

Prior to linking the surface mentions to their respective concepts, named entity recognition [16] is required in order to identify all sequences of tokens in the input sentence that potentially denote an entity of a particular type (e.g., diseases or chemicals). Until recently, named entity recognition and entity linking have been mostly performed as separate tasks in pipeline architectures ([6,19], inter alia).

Although linear chain CRFs are widely used for NEL, recent research outlines the positive impact of complex dependencies between hidden variables that exceeds the limitations of a linear model. We frame the entity recognition and linking tasks as a joint inference problem in a general CRF model. In the following, we describe (i) the underlying factor graph, (ii) the joint inference procedure and (iii) the factor template / feature generation to provide sufficient statistics.

We train and evaluate our system in two experiments focusing on both diseases and chemical compounds, respectively. In both tasks, the *BioCreative V CDR* dataset [24] is used for training and testing. We apply the same model to both domains by only exchanging the underlying reference knowledge base. We show that the suggested model architecture provides high performance on both domains without major need of manual adaptation or system tuning.

3.1 Entity Linking Model and Factor Graph Structure

We define a document as a tuple $d = \langle \boldsymbol{x}, \boldsymbol{m}, \boldsymbol{c}, \boldsymbol{s} \rangle$ comprising an observed sequence of tokens $\boldsymbol{x}$, a set of non-overlapping segments determining entity mentions $\boldsymbol{m}$ and corresponding concepts $\boldsymbol{c}$. We capture possible word synonyms $\boldsymbol{s}$ as hidden variables of individual tokens. In the following, we refer to an annotation $a_i = \langle m_i, c_i, s_i \rangle \in d$ as a tuple of corresponding variables. Further, we define a state as a specific assignment of values to each hidden variable in d. The factor graph of our model is shown in Fig. 4. It consists of hidden variables m, c, and s and observed variables x as well as factor types Ψ_i connecting subsets of these variables. Note that the figure does not show an unrolled factor graph but a general viewpoint to illustrate different types of factors (cf. Fig. 5 for an unrolled example). We distinguish 5 factor types by their instantiating factor template $\{T^1, T^2, T^3, T^4, T^5\} \in \mathcal{T}$ e.g. $\Psi_1 : T^1$ is a factor type that solely connects variables of m.

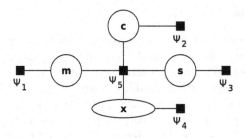

Fig. 4. General factor graph of our model for joint entity recognition and linking. The factor graph consists of hidden variables m, c, and s and observed variables x as well as factor types Ψ^i connecting subsets of these variables.

Let $y = A$ be represented as a set of annotations of the document, then the conditional probability $p(y|x)$ from formula (5) can be written as:

$$P(y|x) = \frac{1}{Z(x)} \overset{M_y}{\underset{m_i}{\prod}} \Psi_1(m_i) \cdot \overset{C_y}{\underset{c_i}{\prod}} \Psi_2(c_i) \cdot \overset{S_y}{\underset{s_i}{\prod}} \Psi_3(s_i) \cdot \overset{X_y}{\underset{x_i}{\prod}} \Psi_4(x_i) \cdot \overset{A_y}{\underset{a_i}{\prod}} \Psi_5(a_i) \quad (15)$$

Factors are formulated as $\Psi_i(\cdot) = \exp(\langle f_{T_i}(\cdot), \theta_{T_i} \rangle)$ with sufficient statistics $f_{T_i}(\cdot)$ and parameters θ_{T_i}. In order to get a better understanding of our model, we illustrate an unrolled version of the factor graph in Fig. 5. Given this example, d can be explicitly written out as: $c = \{c1 = D011507, c_2 = D011507, c_3 = D007674\}$, $s = \{s_3 = disease \rightarrow dysfunction\}$, and $m = \{m_1 = \{x_7\}, m_2 = \{x_{13}\}, m_3 = \{x_{16}, x_{17}\}\}$ and $x = \{x_0, \ldots, x_{17}\}$.

3.2   Inference

Exploring the Search Space. Our inference procedure is based on the MCMC method with the exhaustive Gibbs sampling as defined in Sect. 2.1. The inference procedure is initialized with an empty state s_0 that contains no assignment to any hidden variables, thus $s_0 = \{x = \{x_0, \ldots, x_{n-1}\}, m = \emptyset, s = \emptyset, c = \emptyset\}$. In each iteration, a segmentation-explorer and a concept-explorer are consecutively applied in order to generate a set of proposal states. The segmentation explorer

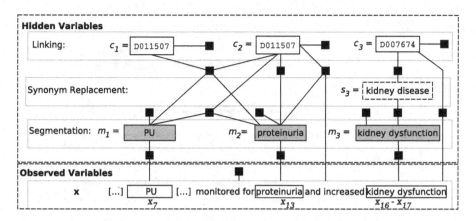

Fig. 5. Unrolled factor graph of our model from Fig. 4 given a concrete example annotated document.

(recognition) is able to add a new non-overlapping segmentation[7], remove an existing segmentation, or apply a synonym replacement to a token within an existing segmentation. The concept-explorer (linking) can assign, change or remove a concept to/from any segmentation.

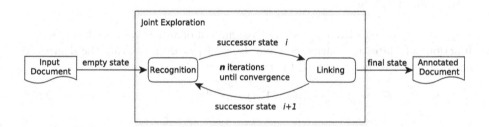

Fig. 6. Illustration of the joint inference procedure for named entity recognition and linking. The procedure begins with an empty state that is passed to the recognition explorer. The successor state is stochastically drawn from the model distribution of proposal states and passed to the linking explorer. We do this for n iterations until convergence.

Applying these explorers in an alternating consecutive manner, as illustrated in Fig. 6, effectively guarantees that all variable assignments are mutually guided by several sources of information: (i) possible concept assignments can inform the segmentation explorer in proposing valid spans over observed input tokens, while (ii) proposing different segmentations together with synonym replacements

---

[7] We do not extend or shrink existing spans. Instead, new annotations can be of different length, spanning 1 to 10 tokens.

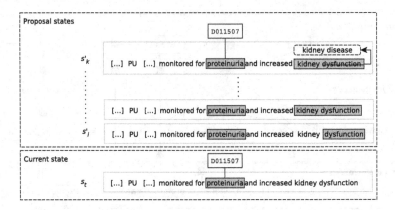

Fig. 7. Subset of proposal states generated by the segmentation explorer, originating from the current state s_t which has already one linked segmentation on token t_{13}. Each proposal state has a new non-overlapping segment annotation (marked in grey) that is not linked to any concept. Proposal states may include synonym replacements (depicted as dashed boxes) that are accepted for all subsequent sampling steps.

on these may facilitate concept linking. Thus, this intertwined sampling strategy effectively enables joint inference on the recognition and the linking task. Figure 7 shows an exemplary subset of proposal states that are generated by the segmentation explorer.

Objective Function. Given a predicted assignment of annotations $\boldsymbol{y}'$ the objective function calculates the harmonic mean based F_1 score indicating the degree of accordance with the ground truth $\boldsymbol{y}^{gold}$. Thus:

$$\mathbb{O} = F_1(\boldsymbol{y}', \boldsymbol{y}^{gold}) \tag{16}$$

3.3 Sufficient Statistics

In the following, we describe our way of creating sufficient statistics by features that encode whether a segmentation and its concept assignment is reasonable or not. All described features are of boolean type and are learned from a set of labeled documents from the training data. We introduce δ as a given dictionary that contains entity surface forms which are linked to concepts, and the bidirectional synonym lexicon κ that contains single token synonyms of the form $x \leftrightarrow x^{synonym}$.

Dictionary Generation

Dictionary Generation. A main component of this approach is a dictionary $\delta \subseteq S \times C$, where $C = \{c_0, \ldots, c_n\}$ is the set of concepts from a reference knowledge base and $S = \{s_0, \ldots, s_m\}$ denotes the set of surface forms that can be used to refer to these concepts. We define two functions on the dictionary: (i) $\delta(s) = \{c \mid (s,c) \in \delta\}$ returns a set of concepts for a given name s, and (ii) $\delta(c) = \{s \mid (s,c) \in \delta\}$ returns a set of names for a given concept c.

Synonym Extraction. We extract a bidirectional synonym lexicon from the dictionary δ by considering all surface forms of a concept c that differ in one token. We consider these tokens as synonyms. For example, the names *kidney disease* and *kidney dysfunction* are names for the same concept and differ in the tokens 'disease' and 'dysfunction'. The replacement (*disease* $\leftrightarrow$ *dysfunction*) is (bidirectional) inserted into the synonym lexicon denoted as κ provided that the pair occurs in at least two concepts.

Feature Generation. For simplicity reasons, we refer in the following with m_i to the underlying text of the ith segmentation and s_i to the underlying text of the corresponding segmentation that includes its synonym replacement. The feature description is guided by the following example sentence:

" $\boxed{\text{Hussein Obama}}$ is the $\boxed{\text{former}}$ president of the $\boxed{\text{USA}}$. "

Here, three segments are annotated (framed tokens). Throughout the concrete feature examples that are provided to each feature description, we denote:

$$m_0 = \text{``Hussein Obama''}, \; s_0 = \{Hussein \leftrightarrow Barack\}, \; c_0 = \varnothing$$
$$m_1 = \text{``former''}, \; s_1 = \varnothing, \; c_1 = \varnothing,$$
$$m_2 = \text{``USA''}, \; s_2 = \varnothing, \; c_2 = dbpedia : United\_States.$$

Dictionary Lookup. For each segmentation m_i in the document, a feature $f^{m_i}_{m_i \in \delta}(y_i)$ is created that indicates whether the text within m_i corresponds to any entry in the dictionary δ. Further, a feature $f^{c_i}_{(m_i,c_i) \in \delta}(y_i)$ indicates whether the text of a segmentation refers to its assigned concept c_i. Analogously, a pair of features is computed that indicate whether s_i is in or is related to the concept c_i according to the dictionary.

$$f^{m_i}_{m_i \in \delta}(y_i) = \begin{cases} 1 & \text{iff } \exists c \in C(m_i, c) \in \delta \\ 0 & \text{otherwise.} \end{cases} \qquad f^{c_i}_{(m_i,c_i) \in \delta}(y_i) = \begin{cases} 1 & \text{iff } (m_i, c_i) \in \delta \\ 0 & \text{otherwise.} \end{cases}$$
$$(17)$$

Example 3

$$f^{m_0}_{m_0 \in \delta}(y_0) = \text{``Hussein Obama''} \in \delta = 1$$

$$f^{c_0}_{(m_0,c_0) \in \delta}(y_0) = (\text{``Hussein Obama''}, \varnothing) \notin \delta = 0$$

$$f^{m_1}_{m_1 \in \delta}(y_1) = \text{``former''} \notin \delta = 0$$

$$f^{c_1}_{(m_1,c_1) \in \delta}(y_1) = (\text{``former''}, \varnothing) \notin \delta = 0$$

$$f^{m_2}_{m_2 \in \delta}(y_2) = \text{``USA''} \in \delta = 1$$

$$f^{c_2}_{(m_2,c_2) \in \delta}(y_2) = (\text{``USA''}, dbpedia : United\_States) \in \delta = 1$$

In this example, *Hussein Obama* and *USA* are part of the dictionary, whereas *former* is not. Further, the assigned concept c_2 to m_2 matches an entry in the dictionary.

Synonyms. Recall that the synonym lexicon κ is generated automatically from training data. Thus, not all entries are meaningful or equally likely and may be concept dependent. Thus, we add a feature f_κ that measures the correlation for a segment m_i to its synonym $s_i \in \kappa$ if any.

$$f^{m_i,s_i}_\kappa(y_i) = \begin{cases} 1 & \text{iff } (m_i, s_i) \in \kappa \\ 0 & \text{otherwise.} \end{cases} \tag{18}$$

Example 4

$$f^{m_0,s_0}_\kappa(y_0) = (\text{``Hussein''} \leftrightarrow \text{``Obama''}) \in \kappa = 1$$

$$f^{m_1,s_1}_\kappa(y_1) = (\text{``former''} \leftrightarrow \varnothing) \notin \kappa = 0$$

$$f^{m_2,s_2}_\kappa(y_2) = (\text{``USA''} \leftrightarrow \varnothing) \notin \kappa = 0$$

In this example, *Hussein Obama* is a synonym for *Barack Obama* based on the synonym lexicon.[8]

Token Length. Given a segment m_i, we consider its length $n_i = len(m_i)$ by binning n_i into discrete values ranging from 1 to n_i: $B = [b_0 = 1, b_1 = 2, \ldots, b_{n-1} = n_i]$. For each element in B, we add a feature f_{len} that tells whether $b_j \in B$ is less or equal to n_i. Analogously, the feature is conjoined with the annotated concept c_i.

$$f^{b_j,n_i}_{len}(y_i) = \begin{cases} 1 & \text{iff } b_j <= n_i \\ 0 & \text{otherwise.} \end{cases} \tag{19}$$

Example 5

$$f^{b_0,n_0}_{len}(y_i) = \text{``len } (1 \leq 2\text{''}) = 1$$

$$f^{b_1,n_0}_{len}(y_i) = \text{``len } (2 \leq 2\text{''}) = 1$$

$$f^{b_0,n_1}_{len}(y_i) = \text{``len } (1 \leq 1\text{''}) = 1$$

$$f^{b_0,n_2}_{len}(y_i) = \text{``len } (1 \leq 1\text{''}) = 1$$

$$f^{b_0,n_2}_{len}(y_i) = \text{``len} + dbpedia : United\_States (1 \leq 1)\text{''} = 1$$

In this example, $n_0 = len(\text{``Barack Obama''}) = 2$, $n_1 = len(\text{``former''}) = 1$, $n_i = len(\text{``USA''}) = 1$.

---

[8] Note that, just because the feature is active it does not mean its a good replacement. This is determined during training.

Token Context and Prior. We capture the context of a segmentation m_i in form of token based $\mathcal{N}$-grams. Let π_k be the kth n-gram within or in the context of m_i, then features of type $f^{m_i,x}_{\pi_k\ context}(y_i)$ and $f^{m_i}_{within}(y_i)$ are created for each π_k that indicate whether a segmentation is (i) preceded by a certain π_k, (ii) followed by π_k, (iii) surrounded by π_k, and (iv) within m_i. In order to model recognition and linking jointly, each of these features is additionally conjoined with the corresponding concept c_i that is: $f^{m_i,c_ix}_{\pi_k\ context}(y_i)$ and $f^{m_i,c_i}_{within}(y_i)$.

Example 6

$$\forall \pi_k \in \Pi_0^{context} : f^{m_0,x}_{\pi_k\ context}(y_0) = \pi_k = 1$$

$$\forall \pi_k \in \Pi_0^{within} : f^{m_0}_{\pi_k\ within}(y_0) = \pi_k = 1$$

$$\forall \pi_k \in \Pi_1^{context} : f^{m_1,x}_{\pi_k\ context}(y_1) = \pi_k = 1$$

$$\forall \pi_k \in \Pi_1^{within} : f^{m_1}_{\pi_k\ within}(y_1) = \pi_k = 1$$

$$\forall \pi_k \in \Pi_2^{context} : f^{m_2,x}_{\pi_k\ context}(y_2) = \pi_k = 1$$

$$\forall \pi_k \in \Pi_2^{within} : f^{m_2}_{\pi_k\ within}(y_2) = \pi_k = 1$$

$$\forall \pi_k \in \Pi_2^{context} : f^{m_2,c_2,x}_{\pi_k\ context}(y_2) = \pi_k + dbpedia:United\_States = 1$$

$$\forall \pi_k \in \Pi_2^{within} : f^{m_2,c_2,x}_{\pi_k\ within}(y_2) = \pi_k + dbpedia:United\_States = 1$$

In this example, we restrict $\mathcal{N}$ to 3 which means we consider only uni-, bi-, and tri-grams. We provide exemplary the $\mathcal{N}$-grams for the first annotation which are: $\Pi_0^{context} = \{$ "is", "the", "former", "is the", "the former", "is the former"$\}$ and $\Pi_0^{within} = \{$ "Hussein", "Hussein Obama", "Obama"$\}$. $\Pi_1^{context}, \Pi_2^{context}$ and $\Pi_1^{within}, \Pi_2^{within}$ are defined analogously.

Coherence. We measure the pairwise coherence of annotations with the feature f_{coh} defined as:

$$f^{a_j,a_k}_{coh}(y_j,y_k) = \begin{cases} 1 & \text{iff } (m_j == m_k) \wedge (s_j == s_k) \wedge (c_j == c_k) \\ 0 & \text{otherwise.} \end{cases} \tag{20}$$

Example 7

$$f^{a_0,a_1}_{coh}(y_0,y_1) = \begin{cases} (\text{"Hussein Obama"} \neq \text{"former"}) \wedge \\ ((\text{"Hussein"} \leftrightarrow \text{"Barack"}) \neq \varnothing) \wedge \quad = 0 \\ (\varnothing == \varnothing) \end{cases}$$

$$f_{coh}^{a_1,a_2}(y_1, y_2) = \begin{cases} (\text{``former''} \neq \text{``USA''}) \land \\ (\varnothing == \varnothing) \land \\ (\varnothing == \varnothing) \end{cases} = 0$$

$$f_{coh}^{a_0,a_2}(y_0, y_2) = \begin{cases} (\text{``Hussein Obama''} \neq \text{``USA''}) \land \\ (\varnothing == \varnothing) \neq \varnothing) \land \\ (\varnothing == \varnothing) \end{cases} = 0$$

For this example, we do not have any active features as they do not share surface forms, concepts and synonym replacements.

Abbreviation. We address the problem of abbreviations (cf. [5]) in the task of entity linking with features f_{abb} that indicate whether the segmentation m_i represents an abbreviation[9] **and** its longform is locally known. That is, iff a non-abbreviation segmentation m_j exists that has the same concept assigned as the abbreviation m_i:

$$f_{abb}^{a_i,a_j}(y_i, y_j) = \begin{cases} 1 & \text{iff } (\text{isAbbr}(m_i) \land \neg\text{isAbbr}(m_j)) \land (c_i == c_j) \land (c_i \neq \varnothing) \\ 0 & \text{otherwise.} \end{cases}$$

$$(21)$$

Example 8

$$f_{abb}^{a_0,a_1}(y_0, y_1) = (false \land true) \land (\varnothing == \varnothing) \land (\varnothing \neq \neg\varnothing) = 0$$

$$f_{abb}^{a_1,a_2}(y_1, y_2) = (false \land true) \land (\varnothing == \varnothing) \land (\varnothing \neq \neg\varnothing) = 0$$

$$f_{abb}^{a_0,a_2}(y_0, y_2) = \begin{cases} (true \land true) \land \\ (\varnothing \neq dbpedia : United\_States) \land \\ (dbpedia : United\_States == \neg\varnothing) \end{cases} = 0$$

For this example, we do not have any active features as no longform of an annotated abbreviation exists that shares the same concept.

3.4 Experiments

The objective of this model is to recognize segments in text denoting an entity of a specific type and link them to a reference knowledge base by assigning a unique concept identifier. In this section, we describe our experiments on two types of biomedical entities. The first experiment evaluates our system in *disease* recognition and linking. The second experiment is conducted on *chemicals*. Both experiments use the same data set described below.

Data Sets and Resources

Data Sets. All experiments were conducted on data from the BioCreative V Shared Task for Chemical Disease Relations (BC5CDR) [24]. The data set was

---

[9] We define an abbreviation as a single token which is in uppercase and has at most 5 characters.

designed to solve the tasks of entity recognition and linking for disease and chemicals and further to find relations between both. However, the latter task is not yet considered in our approach. Each annotation contains information about its span in terms of character offsets and a unique concept identifier. Annotated entities are linked to the Comparative Taxicogenomics Database[10] for diseases (CTD_{dis}) or chemicals (CTD_{chem}), respectively.

The data set consists of 1,500 annotated Pubmed abstracts equally distributed into training, development and test set with about 4,300 unique annotations each.

Reference Knowledge Base. CTD_{dis} is derived from the disease branch of MeSH and the Online Mendelian Inheritance in Man (OMIM)[11] data base. CTD_{dis} contains 11,864 unique disease concept identifiers and 75,883 disease names. CTD_{chem} is solely derived from the chemical branch of MeSH. It comprises 163,362 unique chemical concept identifiers and 366,000 chemical names.

Cleaning Procedure. In order to remove simple spelling variations, we implement a text cleaning procedure which is applied to all textual resources and data sets. The strategy uses six manually created regular expressions like replacing *'s* by *s*. Further, we convert all tokens into lowercase if they are not solely in uppercase, we remove all special characters including punctuation and brackets, and replace multiple whitespace characters by a single blank. We apply the same strategy to both diseases and chemicals.

Resources Used in the Experiments. In the experiments for disease recognition and linking, we initialize the dictionary δ with CTD_{dis} and enhance it with the disease annotations from the training data. We then apply the text cleaning procedure as described above to all entries, as well as to all documents in training and test set. Due to the cleaning, the size of the dictionary reduces to 73,773 unique names ($-2,113$), while the number of concepts remains the same. The resulting synonym lexicon κ stores 2,366 entries.

In the experiments for chemicals, the dictionary δ is initialized with CTD_{chem} and enhanced with the chemical annotations from the training data. After the cleaning procedure, the size of the dictionary reduces to 359,564 unique names (-8.186), while the number of concepts remains the same. The resulting synonym lexicon κ stores 4,912 entries.

The system's overall performance depends on the two parameters k and λ that influence the candidate retrieval procedure (cf. Sect. 3.3), as they determine the maximum recall that can be achieved. We empirically set the best parameter values using a two-dimensional grid search on the development set, assuming perfect entity recognition. Best performance is achieved with $k = 20$ and $\lambda = 0.7$. Given these parameters, a maximum recall of 90.4 for diseases, and 91.5 for chemicals can be obtained by our system on the BC5CDR test set.

---

[10] http://ctdbase.org, version from 2016.
[11] http://www.omim.org.

Baselines. We compare our approach to the two state-of-the-art systems *DNorm* [13] and *TaggerOne* [12], as well as against two simple baselines (*LMB* and *LMB+*). The latter baselines are based on non-overlapping longest matches, using the dictionary as described in Sect. 3.3. While in *LMB+* all resources (including the dictionary and documents) were cleaned, resources in *LMB* remain as they are.

Due to the cleaning, we lose track of the real character offset position. Thus, these baselines are not applicable to the entity recognition subtask.

Experimental Settings

Evaluation Metrics. We use the official evaluation script as provided by the BioCreative V Shared Task organizers [24]. The script uses Precision, Recall and F_1 score on micro level. In the recognition task the measure is on mention level comparing annotation spans including character positions and the annotated text. Experiments on the linking task are evaluated on concept level by comparing *sets* of concepts as predicted by the system and annotated in the gold standard, i.e., multiple occurrences of the same concept and their exact positions in the text are disregarded.

Hyper-parameter Settings. During development, the learning rate α and the number of training epochs ϵ as hyper-parameters of SampleRank were empirically optimized by varying them on the development set. Best results could be achieved with $\alpha = 0.06$. The results reached a stable convergence at $\epsilon = 130$.

Results. We report results on the BC5CDR test set in Table 2. Results on the disease and chemicals subtasks are shown in the left and right part of the table, respectively. For both tasks, we assess the performance of our system on end-to-end entity linking (columns labeled with "Linking"), as well as the entity recognition problem in isolation ("Recognition").

Table 2. Evaluation results on BC5CDR test set for recognition and linking on diseases (left part) and chemicals (right part)

	Diseases						Chemicals					
	Recognition			Linking			Recognition			Linking		
	P	R	F_1	P	R	F_1	P	R	F_1	P	R	F_1
J-Link	84.6	**81.9**	**83.2**	**86.3**	**85.5**	**85.9**	90.0	86.6	88.3	85.9	**91.0**	88.4
TaggerOne	**85.2**	80.2	82.6	84.6	82.7	83.7	**94.2**	**88.8**	**91.4**	88.8	90.3	**89.5**
DNorm	82.0	79.5	80.7	81.2	80.1	80.6	93.2	84.0	88.4	**95.0**	80.8	87.3
LMB+	n/a	n/a	n/a	80.5	80.9	80.7	n/a	n/a	n/a	80.4	82.7	81.5
LMB	n/a	n/a	n/a	82.3	58.5	68.3	n/a	n/a	n/a	84.0	58.8	69.2

Disease Recognition and Linking. In disease recognition, our approach exhibits the best F_1 score of all systems compared here ($F_1 = 83.2$). Only in terms of Precision, TaggerOne has slight advantages.

In the linking task, our system (*J-Link*) clearly outperforms both lexicon-based baselines as well as both state-of-the-art systems. In particular, J-Link exceeds TaggerOne by 2.2 and DNorm by 5.3 points in F_1 score, respectively.

Comparing these results to the baselines, we observe that a simple lexicon lookup (LMB) already achieves robust precision levels that cannot be met by the DNorm system. More than 22 points in recall can be gained by simply applying a cleaning step to the dictionary and documents (LMB$^+$).

However, the increasing recall comes with a drop in precision of 1.8 points. This shows that preprocessing the investigated data can be helpful to find more diseases, while aggravating the linking task. Obviously, our system (in contrast to DNorm and to a greater extent than TaggerOne) benefits from a number of features that provide strong generalization capacities beyond mere lexicon matching.

Chemicals Recognition and Linking. In the second experiment, we are interested in assessing the domain adaptivity of our model. Therefore, we apply the same factor model to a different reference knowledge base, without changing any system parameters or engineering any additional domain-specific features.

The evaluation (cf. Table 2, right part) shows promising results regarding the adaptation to chemicals, particularly in the linking task. Our approach is competitive to DNorm and TaggerOne, while clearly outperforming both lexicon baselines.

Compared to DNorm, our approach lacks in precision (-9.1), but shows better results in recall ($+10.2$), which results in a slightly higher F_1 score ($+1.1$). Overall, TaggerOne obtains the best performance in this experiment, due to the best precision/recall trade-off. However, the superior recall of our system is remarkable (R=91.0), given that the dictionary for chemicals as used in TaggerOne was augmented in order to ensure that all chemical element names and symbols are included [12].

4   Conditional Random Fields for Slot Filling

Initiated by the advent of the distant supervision [15] and open information extraction paradigms [1], the last decade has seen a tendency to reduce information extraction problems to relation extraction tasks. In the latter, the focus is on extracting binary entity-pair relations from text by applying various types of discriminative classification approaches.

We argue that many tasks in information extraction (in particular, when being used as an upstream process for knowledge base population) go beyond the binary classification of whether a given text expresses a given relation or not, as they require the population of complex *template structures*.

We frame template-based information extraction as an instance of a structured prediction problem [22] which we model in terms of a joint probability distribution over value assignments to each of the slots in a template. Subsequently, we will refer to such templates as *schemata* in order to avoid ambiguities with factor templates from the factor graph. Formally, a schema S consists of typed slots $(s_1, s_2, \ldots, s_n)$. The slot-filling task corresponds to the maximum a posteriori estimation of a joint distribution of slot fillers given a document d

$$(s_1, s_2, \ldots, s_n) = \operatorname*{argmax}_{s'_1, s'_2, \ldots, s'_n \in \Phi} P(s_1 = s'_1, \ldots, s_n = s'_n \mid d), \tag{22}$$

where Φ is the set of all possible slot assignments.

Slots in a schema are interdependent, and these dependencies need to be taken into account to avoid incompatible slot assignments. A simple formulation in terms of n binary-relation extraction tasks would therefore be oversimplifying. On the contrary, measuring the dependencies between all slots would render inference and learning intractable. We therefore opt for an intermediate solution, in which we analyze how far measuring *pairwise* slot dependencies helps in avoiding incompatibilities and finally to improve an information extraction model for the task.

We propose a factor graph approach to schema/template-based information extraction which incorporates factors that are explicitly designed to encode such constraints. Our main research interest is therefore to (1) understand whether such constraints can be learned from training data (to avoid the need for manual formulation by domain experts), and (2) to assess the impact of these constraints on the performance.

We evaluate our information extraction model on a corpus of scientific publications reporting the outcomes of pre-clinical studies in the domain of spinal cord injury. The goal is to instantiate multiple schemata to capture the main parameters of each study. We show that both types of constraints are effective, as they enable the model to outperform a naive baseline that applies frequency-based filler selection for each slot.

4.1 Slot Filling Model and Factor Graph Structure

We frame the slot filling task as a joint inference problem in undirected probabilistic graphical models in a distant supervised fashion. Our model is a factor graph which probabilistically measures the compatibility of a given textual document d consisting of tokenized sentences χ, a fixed set of entity annotations $\mathcal{A}$, and a to be filled ontological schema S. The schema S is automatically derived from an ontology and is described by a set of typed slots, $S = \{s_1, \ldots, s_n\}$. Let $\mathcal{C}$ denote the set of all entities from the ontology, then each slot $s_i \in S$ can be filled by a pre-defined subset of $\mathcal{C}$ called slot filler. Further, each annotation $a \in \mathcal{A}$ describes a tuple $\langle t, c \rangle$ where $t = (t_i, \ldots, t_j) \in \chi$ is a sequence of tokens with length ≥ 1 and a corresponding filler type $c \in \mathcal{C}$.

Factorization of the Probability Distribution. We decompose the overall probability of a schema S into probability distributions over single slot and pairwise slot fillers. Each individual probability distribution is described through factors that measure the compatibility of single/pairwise slot assignments. An unrolled factor graph that represents our model structure is depicted in Fig. 8. The factor graph consists of different types of factors that are connected to subsets of variables of $\boldsymbol{y} = \{y_0, y_1, \ldots, y_n\}$ and of $\boldsymbol{x} = \{\chi, \mathcal{A}\}$, respectively. We distinguish three factor types by their instantiating factor template $\{T', T'_d, T''_d\} \in \mathcal{T}$: (i) **Single slot factors** $\Psi'(y_i) \in T'$ that are solely connected to a single slot y_i, (ii) **Single slot+text factors** $\Psi'(y_i, \boldsymbol{x}) \in T'_d$ that are connected to a single slot y_i and $\boldsymbol{x}$, (iii) **Pairwise slot+text factors** $\Psi''(y_i, y_j, \boldsymbol{x}) \in T''_d$ that are connected to a pair of two slots y_i, y_j and $\boldsymbol{x}$.

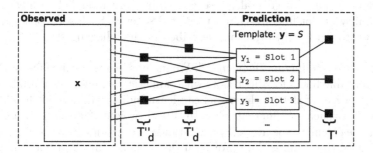

Fig. 8. Factor graph of our model for an exemplary ontological schema S. It shows three different types of factors. Each set of factors of the same type is instantiated by a different factor template.

The conditional probability $P(\boldsymbol{y} \mid \boldsymbol{x})$ of a slot assignment $\boldsymbol{y}$ given $\boldsymbol{x}$ can be simplified as:

$$P(\boldsymbol{y}|\boldsymbol{x}) = \frac{1}{Z(\boldsymbol{x})} \prod_{y_i \in S} \left[\Psi'(y_i) \cdot \Psi'(y_i, \boldsymbol{x}) \right] \prod_{y_i \in S} \prod_{y_j \in S} \left[\Psi''(y_i, y_j, \boldsymbol{x}) \right]. \tag{23}$$

Factors are formulated as $\Psi(\cdot) = \exp(\langle f_T(\cdot), \theta_T \rangle)$ with sufficient statistics $f_T(\cdot)$ and parameters θ_T ($T \in \mathcal{T}$ and $\Psi \in \{\Psi', \Psi''\}$).

4.2 Inference and Learning

Ontological Sampling. The initial state s_0 in our exploration is empty, thus $\boldsymbol{y} = (\varnothing)$. A set of potential successors is generated by a proposal function changing a slot by either deleting an already assigned value or changing the value to another slot filler. The successor state s_{t+1} is chosen based on the probability distribution generated by the model. The higher the probability (according to the model) of a state, the higher is the chance of being chosen as successor state. However, the state is only accepted iff $q(s_{t+1}) > q(s_t)$, where $q(s')$ is the model probability of the state s'. The inference procedure stops if the state selected for each sampling step does not change for three iterations.

Objective Function. Given a predicted assignment $\boldsymbol{y}^*$ of all slots in schema type $\hat{S}$ and a set $\mathcal{S}$ of instantiated schemata of type $\hat{S}$ from the gold standard, the training objective is

$$\mathbb{O}(\boldsymbol{y}^*) = \max_{\boldsymbol{y}' \in \mathcal{S}} F_1(\boldsymbol{y}^*, \boldsymbol{y}'), \tag{24}$$

where F_1 is the harmonic mean of precision and recall, based on the overlap of assigned slot values between $\boldsymbol{y}$ and $\boldsymbol{y}'$.

4.3   Factors and Constraints

At the core of this model are features that encode soft constraints to be learned from training data. In general, these constraints are intended to measure the compatibility of slot fillers within a predicted schema. Such soft constraints are designed through features that are described in the following.

Single-Slot Constraints in Template T'. We include features which measure common, acceptable fillers for single slots with numerical values. Given a filler annotation $a_i = \langle v, c \rangle$ of slot y_i, the model can learn individual intervals for different types of fillers such as temperature $(-10\text{--}40)$, or weight $(200\text{--}500)$, for example. For that, we calculate the average μ and standard deviation σ for each particular slot based on the training data. For each slot s_i in schema S, a boolean feature $f^{s_i}_{\sigma=n}$ is instantiated for each $n \in \{0, \ldots, 4\}$, indicating whether the value y_i is within n standard deviations σ_{s_i} of the corresponding mean μ_{s_i}. To capture the negative counterpart, a boolean feature $f^{s_i}_{\sigma>n}$ is instantiated likewise.

$$f^{s_i}_{\sigma=n}(y_i) = \begin{cases} 1 & \text{iff } \lceil (\frac{v-\mu_{s_i}}{\sigma_{s_i}}) \rceil = n \\ 0 & \text{otherwise.} \end{cases} \qquad f^{s_i}_{\sigma>n}(y_i) = \begin{cases} 1 & \text{iff } \lceil (\frac{v-\mu_{s_i}}{\sigma_{s_i}}) \rceil > n \\ 0 & \text{otherwise.} \end{cases} \tag{25}$$

In this way, the model learns preferences over possible fillers for a given slot which effectively encode soft constraints such as "the weight of rats typically scatters around a mean of 300 g by two standard deviations of 45 g".

Pairwise Slot Constraints in T''_d. In contrast to single-slot constraints, pairwise constraints are not limited to slots with filler type $v \in \mathbb{R}$. Soft constraints on slot pairs are designed to measure the compatibility and (hidden) dependencies between two fillers, e.g., the dependency between the dosage of a medication and its applied compound, or between the gender of an animal and its weight. This is modeled in terms of their linguistic context and textual locality, as discussed in the following.

We assume that possible slot fillers may be mentioned multiple times at various positions in a text. Therefore, given a pair of slots (s_i, s_j), we define λ as an aggregation function that returns the subset of annotations $\lambda(s_i) = \{a = \langle t, c \rangle \in \mathcal{A} \mid a(c) = s_i(c)\}$. We measure the locality of two slots in the text by the minimum distance between two sentences containing annotations for the corresponding slot fillers. A bi-directional distance for two annotations is defined as $\delta(a_k, a_l) = |\text{sen}(a_k) - \text{sen}(a_l)|$ where *sen* denotes a function that returns the sentence index of an annotation. For each $n \in \{0, \ldots, 9\}$ a boolean feature $f_{\delta=n}$ is instantiated as:

$$f_{\delta=n}^{s_i,s_j}(y_i, y_j) = \begin{cases} 1 & \text{iff } n = \min_{a_k \in \lambda(y_i), a_l \in \lambda(y_j)} \delta(a_k, a_l) \\ 0 & \text{otherwise.} \end{cases} \tag{26}$$

To capture the linguistic context between two slot fillers y_i and y_j, we define a feature $f_{\pi_n}^{s_i}(y_i, y_j)$ that indicates whether a given $\mathcal{N}$-gram $\pi_n \in \pi$ with $1 < \mathcal{N} \leq 3$ occurs between the annotations $a_k \in \lambda(y_i)$ and $a_l \in \lambda(y_i)$ in the document.

Standard Textual Features in T' and T_d'. Given a single slot s_i with filler y_i and the aggregated set of all corresponding annotations $\lambda(y_i)$, we instantiate three boolean features for each annotation $a \in \lambda(y_i)$ as follows.

Let $L_s(l_{y_i}, a(t))$ be the Levenshtein similarity between the ontological class label l_{y_i}, and the tokens of an annotation $a(t)$. Two boolean features $f_{bin(s_{max})<\Delta}(y_i)$ and $f_{bin(s_{max}) \geq \Delta}(y_i)$ are computed as:

$$f_{bin(s_{max})<\Delta}(y_i) = \begin{cases} 1 & \text{iff } b < \Delta \\ 0 & \text{otherwise.} \end{cases} \quad f_{bin(s_{max}) \geq \Delta}(y_i) = \begin{cases} 1 & \text{iff } b \geq \Delta \\ 0 & \text{otherwise.} \end{cases}, \tag{27}$$

where $b = bin(s_{max})$ is the discretization of the maximum similarity s_{max} into intervals of size 0.1, and

$$s_{max} = \max_{a \in \lambda(y_i)} L_s(l_{y_i}, a(t)) \text{ with } L_s = 1 - \frac{levenshtein(l_{y_i}, a(t))}{\max(len(l_{y_i}), len(a(t)))}. \tag{28}$$

Finally, we instantiate features $f_{\pi_k\ context}^{s_i}(s_i)$ and $f_{within}^{s_i}$, indicating whether an $\mathcal{N}$-gram π_k occurs in the context (before or after) or within any annotation of slot y_i.

4.4 Cold-Start Knowledge Base Population in the Spinal Cord Injury Domain

Problem Description. We address the problem of ontology-based information extraction in a slot-filling setting as a prerequisite for cold-start knowledge base population. The extraction task comprises multiple schemata of different types, each of them provided by a domain ontology and containing multiple slots. Each slot in a schema needs to be filled either by a literal from the input document or by a class or individual from the ontology, depending on whether it is derived from a data-type or object-type property

We consider slot-filling as a document-level task, i.e., entities filling the slots of a particular schema may be dispersed across the entire text. In addition, each literal or ontological category can, in principle, fill multiple slots of the appropriate type. We approach the task in a supervised machine learning approach; supervision is available at the document level in terms of fully instantiated gold schemata without direct links between slot fillers and text mentions.

Spinal Cord Injury Ontology (SCIO). Pre-clinical trials in the spinal cord injury domain follow strict methodological patterns. Experimental protocols and the main outcomes of pre-clinical studies on spinal cord injury are formally represented in SCIO [2]. In total, the ontology contains more than 500 classes and approx. 80 properties (slots). SCIO top-level classes defining the schema types are ANIMALMODEL, INJURYMODEL, TREATMENT, INVESTIGATIONMETHOD and RESULT. Slots are either object-type properties which can be filled by a SCIO class, or data-type properties which are filled with free text.

Annotated Data Set. The annotated data set was created by two SCI experts who annotated 25 full-text scientific papers from the SCI literature. Annotations were provided at the level of fully instantiated schemata per document, using the set of top-level classes in SCIO and their corresponding properties as annotation schema. The entire annotation process comprises three steps: (i) mention identification, (ii) entity recognition (in case of data-type properties) and linking (object-type properties), (iii) schema instantiation, and (iv) filling the slots of an instantiated schema with an appropriate entity. The latter steps are due to the fact that the cardinality of schemata of a particular type per document is unknown a priori, and multiple schemata may share individual slot fillers. The following example shows a sentence that describes two instantiations of an ANIMALMODEL schema which share the slot fillers *species* (SPRAGUEDAWLEYRAT) and *ageCategory* (ADULT): *"A total of 39 Sprague-Dawley rats were used for these experiments: adult males (285-330 g) and females (192-268 g)."*

Inter-annotator agreement at the level of fully instantiated schemata in terms of F_1 score between annotators amounts to 0.93 for ANIMALMODEL, 0.79 for INJURY, 0.77 for TREATMENT and 0.65 for INVESTIGATIONMETHOD.

4.5   Experiments

In the following section, we describe our experimental settings, the evaluation metrics and results. Model performances are independently reported for four SCIO schemata: ANIMALMODEL, INJURY, TREATMENT, and INVESTIGATIONMETHOD. As a preprocessing step, we apply symbolic entity recognition in order to generate annotations $\mathcal{A}$. The regular expressions used are automatically generated from ontology class labels. In case of data-type properties (e.g., weight of an animal), regular expressions are manually created.

Experimental Settings. The system is evaluated in a 6-fold cross validation on the complete data set. In all experiments, we restrict the complexity of the schemata to first-order slots, i.e., ontological properties that are directly connected to their respective domain class. In the current approach, we are not aiming at predicting the correct number of instantiations per schema type. Thus, our system is restricted to fill a single schema of each type per document, even if it contains multiple instances of the same schema type (e.g., multiple TREATMENTs).

With respect to this restriction, we report the evaluation results for both, (i) *Full Evaluation* (taking the actual number of gold schemata into account), and (ii) *Best Match Evaluation* (comparing the predicted schema to the best matching gold schema).

Further, we report the performance for two different models, in order to investigate the relative impact of single-slot constraints vs. pairwise slot constraints. In the *pairwise slot filling* (PSF) model, the inference and the factor graph is based on the joint assignment of slot pairs, whereas in *single slot filling* (SSF) model, all slots are independently filled.

Evaluation Metrics. We report model performances as macro precision, recall and harmonic F_1. Given a document with a set of gold schemata $\mathcal{G}$ of type $S = \{s_0, \ldots s_n\}$ and the predicted schema p, the comparison is always based on the best assignment $g' = \mathrm{argmax}_{g \in \mathcal{G}} F_1(p, g)$. For the computation of the overall F_1 score, we convert all ontological schemata into sets of slot-filler pairs with $p = \{s'_0 = c_j, \ldots, s'_n = c_k\}$ and $\mathcal{G} = \{g^0, \ldots, g', \ldots, g^l\} = \{(s^0_0 = c_a, \ldots, s^0_n = c_b), \ldots, (s'_0 = c_c, \ldots, s'_n = c_d), \ldots, (s^l_0 = c_e, \ldots, s^l_n = c_f)\}$. The overall F_1 score is calculated based on the two sets of p and $\mathcal{G}$. We define a true positive (tp) as a slot-filler pair that are in both p and $\mathcal{G}$, a false positive (fp) as a pair that is in p but not in $\mathcal{G}$, and a false negative (fn) as a pair that is in $\mathcal{G}$ but not in p. During the *Best Match Evaluation*, we set $\mathcal{G} = \{g'\}$.

Most Frequent Filler Baseline. We compare the performance of our models in all settings against a plausible but naive baseline. Following the intuition that important information is mentioned in a higher frequency than non-important information, a slot is always filled with the filler that has the highest annotation frequency. In the following, we refer to this procedure as Most Frequent Filler (MFF) baseline.

Results. In the following, we describe the evaluation results for all experiments. First, we compare the performance in the *Full Evaluation* vs. *Best Match Evaluation* settings. In the former setting, we expect a rather low recall due to the restriction of predicting exactly one schema per type. This leads to many false negatives, as multiple instances of the same type can not be fully covered yet. Hence, we hypothesize a significant increase in recall in the *Best Match Evaluation* setting. By comparing the predicted schema to the best match only, we investigate whether the low recall is due to the large amount of missing schemata. If so, this would indicate that our model is able to select the correct slot fillers among a huge set of possible candidates. The performance of all models in both settings is reported in Table 3.

Full Evaluation Results. The results show a strong recall of our baseline model with a distinct lack in precision. The baseline yields the highest recall among all models and schema types except for the ANIMALMODEL (0.55 for baseline vs. 0.90 for SSF/PSF). Compared to the SSF model, we notice a considerable increase in precision in all schema types which is most pronounced in the

Table 3. Performance of Most Frequent Filler Baseline (MFF) vs. Single Slot Filler (SSF) and Pairwise Slot Filler (PSF) models in the *Full Evaluation* (full) and *Best Match* (best) setting.

		MFF			SSF			PSF		
		P	R	F_1	P	R	F_1	P	R	F_1
ANIMAL	full	0.48	0.55	0.51	0.84	0.90	0.86	**0.91**	**0.90**	**0.90**
MODEL	best	0.48	0.57	0.52	0.84	**1.00**	0.91	**0.91**	**1.00**	**0.95**
INJURY	full	0.28	**0.38**	0.31	0.52	0.22	0.31	**0.77**	0.30	**0.43**
	best	0.28	**0.43**	0.33	0.52	0.29	0.35	**0.77**	0.40	**0.50**
TREATMENT	full	0.39	**0.26**	**0.30**	0.70	0.16	0.26	**0.87**	0.16	0.27
	best	0.39	**0.74**	0.51	0.70	0.63	0.65	**0.87**	0.63	**0.73**
INVEST.METHOD	full	0.36	**0.45**	0.36	**1.00**	0.39	0.50	**1.00**	0.39	**0.50**
	best	0.36	0.98	0.52	**1.00**	**1.00**	**1.00**	**1.00**	**1.00**	**1.00**

INVESTIGATIONMETHOD (+0.64). The increase in precision for the three other schemata are between +0.24 and +0.36. Comparing the PSF to the SSF model, we observe further strong improvements in precision and slight improvements in recall. The PSF model clearly outperforms the baseline for the ANIMALMODEL with an increase in F_1 of +0.39, the INJURY +0.12, and the INVESTIGATION-METHOD with +0.14. Despite the precision being increased by +0.46 in the TREATMENT, the baseline shows a higher F_1 score in this configuration (+0.03), due to a drop in recall by −0.10.

Best Match Evaluation Results. In this setting, we further investigate the recall performance of our models compared to the previously discussed *Full Evaluation* results. As we only remove uncaptured schema instances from $\mathcal{G}$ (cf. Sect. 4.5), the precision remains the same. All models show an overall increase in recall for all schema types. With respect to the PSF model, we can see a strong increase in recall for INVESTIGATIONMETHOD by +0.61 and for TREATMENT by +0.47. Further, slight increases by +0.10 and +0.07 can be observed for ANIMALMODEL and INJURY, respectively. Similar observations can be made for the SSF model.

Discussion. Comparing the baseline model with the SSF model, we notice a very strong increase in precision in combination with a slight drop in recall. This positive trend in precision is continued when considering the PSF model. Further, the results show a positive impact of pairwise over single-slot constraints on recall.

The high recall of 0.90 for the ANIMALMODEL in the full evaluation is mainly due to a low number (1 to 2) of instances per schema type in each document. The fact that there is no difference in the performance of the SSF and SPF models for the INVESTIGATIONMETHOD suggests a strong slot independence, so that pairwise slot constraints do not have a big impact. The low increase in recall between the two evaluation settings for the INJURY suggests difficulties for this schema. In contrast, the recall increase for the TREATMENT schema from 0.16 to 0.63 clearly shows that most of the errors are due to a large number of schema instances per document.

Overall, the results show that our system is often able to select the correct set of slot fillers for a schema, even from a huge set of possible schemata and their corresponding slot filler candidates.

5   Conclusion

In this paper accompanying our tutorial, we have discussed how the cold-start knowledge base population task can be modeled as structure-to-structure prediction problems as statistical inference. We have adopted the framework of conditional random fields which represent parametrized conditional probability densities in which a set of output variables is conditioned on a set of input variables. The conditional distribution is represented as a product of so called local factors that model the compatibility between assignments to a subset of variables. The structure of a CRF is typically represented by a factor graph that connects factors to the variables in their scope.

We have shown how tasks in knowledge base population that consist in predicting the most likely instantiation of a given ontology structure given a document can be modeled as statistical inference using conditional random fields. As two prominent problems we have shown how the problem of linking named entities to the corresponding URI representing the real world entity as well as the problem of slot filling can be solved using the proposed framework.

Acknowledgments. This work has been funded by the Federal Ministry of Education and Research (BMBF, Germany) in the PSINK project (project number 031L0028A).

References

1. Banko, M., Cafarella, M., Soderland, S., Broadhead, M., Etzioni, O.: Open information extraction from the web. In: Proceedings of IJCAI, pp. 2670–2676 (2007)
2. Brazda, N., et al.: SCIO: an ontology to support the formalization of pre-clinical spinal cord injury experiments. In: Proceedings of the 3rd JOWO Workshops: Ontologies and Data in the Life Sciences (2017)
3. Freitag, D.: Machine learning for information extraction in informal domains. Mach. Learn. **39**(2–3), 169–202 (2000)
4. Hartung, M., ter Horst, H., Grimm, F., Diekmann, T., Klinger, R., Cimiano, P.: SANTO: a web-based annotation tool for ontology-driven slot filling. In: Proceedings of the 56th Annual Meeting of the Association for Computational Linguistics (System Demonstrations), Association for Computational Linguistics (2018). in press
5. Hartung, M., Klinger, R., Zwick, M., Cimiano, P.: Towards gene recognition from rare and ambiguous abbreviations using a filtering approach. Proc. BioNLP **2014**, 118–127 (2014)

6. Hoffart, J., et al.: Robust disambiguation of named entities in text. In: Proceedings of EMNLP, pp. 782–792 (2011)

7. ter Horst, H., Hartung, M., Cimiano, P.: Joint entity recognition and linking in technical domains using undirected probabilistic graphical models. In: Gracia, J., Bond, F., McCrae, J.P., Buitelaar, P., Chiarcos, C., Hellmann, S. (eds.) LDK 2017. LNCS (LNAI), vol. 10318, pp. 166–180. Springer, Cham (2017). https://doi.org/10.1007/978-3-319-59888-8_15

8. ter Horst, H., Hartung, M., Klinger, R., Brazda, N., Müller, H.W., Cimiano, P.: Assessing the impact of single and pairwise slot constraints in a factor graph model for template-based information extraction. In: Silberztein, M., Atigui, F., Kornyshova, E., Métais, E., Meziane, F. (eds.) NLDB 2018. LNCS, vol. 10859, pp. 179–190. Springer, Cham (2018). https://doi.org/10.1007/978-3-319-91947-8_18

9. Koller, D., Friedman, N.: Probabilistic Graphical Models. Principles and Techniques. MIT Press, Cambridge (2009)

10. Kschischang, F.R., Frey, B.J., Loeliger, H.A.: Factor graphs and sum product algorithm. IEEE Trans. Inf. Theor. $47(2)$, 498–519 (2001)

11. Lafferty, J., McCallum, A., Pereira, F.: Conditional random fields probabilistic models for segmenting and labeling sequence data. In: Proceedings of ICML, pp. 282–289 (2001)

12. Leaman, R., Lu, Z.: TaggerOne Joint named entity recognition and normalization with Semi-Markov Models. Bioinformatics 32, 2839–46 (2016)

13. Leaman, R., Dogan, R.I., Lu, Z.: DNorm disease name normalization with pairwise learning to rank. Bioinformatics 29, 2909–2917 (2013)

14. Min, B., Freedman, M., Meltzer, T.: Probabilistic inference for cold startknowledge base population with prior world knowledge. In: Proceedings of the 15th Conference of the European Chapter of the Association for Computational Linguistics vol. 1, Long Papers, pp. 601–612. Association for Computational Linguistics, Valencia, Spain (April 2017)

15. Mintz, M., Bills, S., Snow, R., Jurafsky, D.: Distant supervision for relation extraction without labeled data. In: Proceedings of ACL, pp. 1003–1011 (2009)

16. Nadeau, D., Sekine, S.: A survey of named entity recognition and classification. Lingvisticae Invest. $30(1)$, 3–26 (2007)

17. Piskorski, J., Yangarber, R.: Information extraction: past, present and future. In: Poibeau, T., Saggion, H., Piskorski, J., Yangarber, R. (eds.) Multi-source Multilingual Information Extraction and Summarization Theory and Applications of Natural Language Processing. Springer, Heidelberg (2013). https://doi.org/10.1007/978-3-642-28569-1_2

18. Poon, H., Domingos, P.: Machine reading: a "Killer App" for statistical relational AI. In: Proceedings of StarAI, pp. 76–81 (2010)

19. Ratinov, L., Roth, D., Downey, D., Anderson, M.: Local and global algorithms for disambiguation to wikipedia. In: Proceedings of ACL:HLT, pp. 1375–1384 (2011)

20. Resnik, P., Hardisty, E.: Gibbs sampling for the uninitiated. Maryland Univ College Park Inst for Advanced Computer Studies, Technical report (2010)

21. Röder, M., Usbeck, R., Ngomo, A.C.N.: Gerbil-benchmarking named entity recognition and linking consistently. Semantic Web J. (2018), http://www.semantic-web-journal.net/system/files/swj1671.pdf

22. Smith, N.A.: Linguistic Structure Prediction. Morgan and Claypool, San Rafael (2011)

23. Sutton, C., McCallum, A.: An introduction to conditional random fields. Foundations and Trends® in Machine Learning $4(4)$, 267–373 (2012)

24. Wei, C.H., et al.: Overview of the biocreative V chemical disease relation (CDR) task. In: Proceedings of the BioCreative V Evaluation Workshop, pp. 154–166 (2015)
25. Wick, M., Rohanimanesh, K., Culotta, A., McCallum, A.: SampleRank learning preferences from atomic gradients. In: Proceedings of the NIPS Workshop on Advances in Ranking, pp. 1–5 (2009)
26. Wimalasuriya, D.C., Dou, D.: Ontology-based information extraction: an introduction and a survey of current approaches. J. Inf. Sci. **36**(3), 306–323 (2010)

Machine Learning with and for Semantic Web Knowledge Graphs

Heiko Paulheim[✉]

Data and Web Science Group, University of Mannheim, Mannheim, Germany
`heiko@informatik.uni-mannheim.de`

Abstract. Large-scale cross-domain knowledge graphs, such as DBpedia or Wikidata, are some of the most popular and widely used datasets of the Semantic Web. In this paper, we introduce some of the most popular knowledge graphs on the Semantic Web. We discuss how machine learning is used to improve those knowledge graphs, and how they can be exploited as background knowledge in popular machine learning tasks, such as recommender systems.

Keywords: Knowledge graphs · Semantic web · Machine learning
Background knowledge

1 Introduction

The term "Knowledge Graph" was coined by Google when they introduced their knowledge graph as a backbone of a new Web search strategy in 2012, i.e., moving from pure text processing to a more symbolic representation of knowledge, using the slogan "things, not strings"[1].

A similar idea, albeit already introduced in the mid-2000s, underlies the concept of *Linked Data*: in order to organize knowledge, URIs (instead of textual names) are used to identify and distinguish entities [1]. Hence, many datasets of the Linked Open Data cloud [73] could also be considered knowledge graphs.

There is no formal definition of a knowledge graph [12]. In the course of this work, we follow the characteristics sketched in [47], saying that a knowledge graph

1. mainly describes real world entities and their interrelations, organized in a graph.
2. defines classes and properties of entities in a schema.
3. allows for potentially interrelating arbitrary entities with each other.
4. covers various topical domains.

We call a dataset following those characteristics and published using Semantic Web standards a *Semantic Web Knowledge Graph*. As Semantic Web standards, we understand

---

[1] https://googleblog.blogspot.de/2012/05/introducing-knowledge-graph-things-not.html.

© Springer Nature Switzerland AG 2018
C. d'Amato and M. Theobald (Eds.): Reasoning Web 2018, LNCS 11078, pp. 110–141, 2018.
https://doi.org/10.1007/978-3-030-00338-8_5

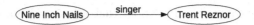

Fig. 1. Example RDF statement

- the use of *derefencable URIs* [1] for referring to entities, i.e., URIs that point to resources on the Web
- the use of RDF[2] for representing the graph, and
- the use of RDF schema[3] and/or OWL[4] for representing the schema of the graph.

RDF organizes knowledge in *statements*, connecting either two entities in a knowledge graph by an edge, or an entity with a literal (i.e., elementary) value (such as a number or a date). Figure 1 shows an example of such an RDF statement. It can also be written down as a *triple* consisting of a subject, a predicate, and an object, where the subject and the object are the entities, whereas the predicate is the property or edge label:

:Nine_Inch_Nails :singer :Trent_Reznor.

Hence, an RDF knowledge graph can either be conceived as a directed, labeled graph, or as a set of such triples.

As an alternative to the triple notation, such statements can be expressed in terms of binary predicates, e.g.,

singer(Nine_Inch_Nails, Trent_Reznor)

In the course of this paper, we will use the following terms to refer to concepts related to knowledge graphs:

entities or *instances* are the nodes in a graph. Typically, they refer to an entity in the real world, such as a person, a city, etc.
literals are elementary data values, such as numbers or dates. They can be used, e.g., for expressing the birth date of a person or the population of a city.
relations are the edges in a graph. They link two entities or an entity and a literal.

Those concepts are typically used in the *A-box*, i.e., the assertional part of the knowledge graph. This is typically the larger part of a knowledge graph. It is complemented by the schema, or *T-box*, which defines the types of entities and relations that can be used in a knowledge graph. Those encompass:

classes or *types* are the categories of entities that exist in a knowledge graph, e.g., *Person, City*, etc. They can form a hierarchy, e.g., *City* being a subclass of *Place*.

---

[2] https://www.w3.org/RDF/.
[3] https://www.w3.org/TR/rdf-schema/.
[4] https://www.w3.org/TR/owl-overview/.

properties are the categories of relations that exist in a knowledge graph, e.g., *birth date*, *birth place*, etc.

While the set of classes is defined in the T-box, the assertion that an individual class is made as a triple in the A-box, e.g.:

:Nine_Inch_Nails a :Band.

or using a unary predicate:

Band(Nine_Inch_Nails)

All assertions made in the A-box – relating two entities, relating an entity and a literal, and assigning a type to an entity, are called *facts*.

Often, the schema also defines further constraints, e.g., certain classes may be disjoint, and properties may have a *domain* (i.e., all subjects of the relation have a certain type) and a *range* (i.e., all objects of the relation have a certain type). For example, the relation *birth Place* may have the domain Person and the range City. Moreover, properties may be defined with other characteristics, such as symmetry, transitivity, etc.

2   Semantic Web Knowledge Graphs

Various public knowledge graphs are available on the Web, including DBpedia [30] and YAGO [33], both of which are created by extracting information from Wikipedia (the latter exploiting WordNet on top), the community edited Wikidata [79], which imports other datasets, e.g., from national libraries[5], as well as from the discontinued Freebase [58], the expert curated OpenCyc [32], and NELL [5], which exploits pattern-based knowledge extraction from a large Web corpus. Lately, new knowledge graphs have been introduced, including DBkWik, which transfers the DBpedia approach to a multitude of Wikis [26], and WebIsALOD, which extracts a knowledge graph of hypernymy relations from the Web [24,74] using Hearst patterns [20] on a large-scale Web corpus.

2.1   Cyc and OpenCyc

The *Cyc* knowledge graph is one of the oldest knowledge graphs, dating back to the 1980s [32]. Rooted in traditional artificial intelligence research, it is a *curated* knowledge graph, developed and maintained by CyCorp Inc.[6] OpenCyc was a reduced version of Cyc, which used to be publicly available, including a Linked Open Data endpoint with links to DBpedia and other LOD datasets. Despite its wide adoption, OpenCyc was shut down in 2017.[7]

OpenCyc contained roughly 120,000 entities and 2.5 million facts; its schema comprised a class hierarchy of roughly 45,000 classes, and 19,000 properties.[8]

---

[5] https://www.wikidata.org/wiki/Wikidata:Data_donation.
[6] http://www.cyc.com/.
[7] http://www.cyc.com/opencyc/.
[8] These numbers have been gathered by own inspections of the 2012 of version of OpenCyc.

2.2 Freebase

Curating a universal knowledge graph is an endeavour which is infeasible for most individuals and organizations. To date, more than 900 person years have been invested in the creation of Cyc [72], with gaps still existing. Thus, distributing that effort on as many shoulders as possible through *crowdsourcing* is a way taken by *Freebase*, a public, editable knowledge graph with schema templates for most kinds of possible entities (i.e., persons, cities, movies, etc.). After MetaWeb, the company running Freebase, was acquired by Google, Freebase was shut down on March 31st, 2015.

The last version of Freebase contains roughly 50 million entities and 3 billion facts[9]. Freebase's schema comprises roughly 27,000 classes and 38,000 relation properties.[10]

2.3 Wikidata

Like Freebase, *Wikidata* is a collaboratively edited knowledge graph, operated by the Wikimedia foundation[11] that also hosts the various language editions of Wikipedia. After the shutdown of Freebase, the data contained in Freebase is subsequently moved to Wikidata.[12] A particularity of Wikidata is that for each axiom, provenance metadata can be included – such as the source and date for the population figure of a city [79].

To date, Wikidata contains roughly 16 million entities[13] and 66 million facts[14]. Its schema defines roughly 23,000 classes[15] and 1,600 properties[16].

2.4 DBpedia

DBpedia is a knowledge graph which is extracted from structured data in Wikipedia. The main source for this extraction are the key-value pairs in the Wikipedia infoboxes. In a crowd-sourced process, types of infoboxes are mapped to the DBpedia ontology, and keys used in those infoboxes are mapped to properties in that ontology. Based on those mappings, a knowledge graph can be extracted [30].

The most recent version of the main DBpedia (i.e., DBpedia 2016-10) contains 5.1 million entities and almost 400 million facts.[17] The ontology comprises 754 classes and 2,849 properties.

---

[9] http://www.freebase.com.
[10] These numbers have been gathered by queries against Freebase's query endpoint.
[11] http://wikimediafoundation.org/.
[12] http://plus.google.com/109936836907132434202/posts/3aYFVNf92A1.
[13] http://www.wikidata.org/wiki/Wikidata:Statistics.
[14] http://tools.wmflabs.org/wikidata-todo/stats.php.
[15] http://tools.wmflabs.org/wikidata-exports/miga/?classes#_cat=Classes.
[16] http://www.wikidata.org/wiki/Special:ListProperties.
[17] http://wiki.dbpedia.org/dbpedia-2016-04-statistics.

2.5 YAGO

Like DBpedia, YAGO is also extracted from DBpedia. YAGO builds its classification implicitly from the category system in Wikipedia and the lexical resource *WordNet* [41], with infobox properties manually mapped to a fixed set of attributes. While DBpedia creates different interlinked knowledge graphs for each language edition of Wikipedia [4], YAGO aims at an automatic fusion of knowledge extracted from various Wikipedia language editions, using different heuristics [33].

The latest release of YAGO, i.e., YAGO3, contains 5.1 million entities and 1.5 billion million facts. The schema comprises roughly 488,000 classes and 77 properties [33].

2.6 NELL

While DBpedia and YAGO use semi-structured content as a base, methods for extracting knowledge graphs from unstructured data have been proposed as well. One of the earliest approaches working at web-scale was the *Never Ending Language Learning (NELL)* project [5]. The project works on a large-scale corpus of web sites and exploits a coupled process which learns text patterns corresponding to type and relation assertions, as well as applies them to extract new entities and relations. Reasoning using a light-weight ontology[18] is applied for consistency checking and removing inconsistent axioms. The system is still running today, continuously extending its knowledge base. While not published using Semantic Web standards, it has been shown that the data in NELL can be transformed to RDF and provided as Linked Open Data as well [83].

In its most recent version, NELL contains roughly 2 million entities and 3 million relations between those. The NELL ontology defines 285 classes and 425 properties.

2.7 DBkWik

DBkWik uses the software that is used to build DBpedia, and applies it to a multitude of Wiki dumps collected from a large Wikifarm, i.e., *Fandom powered by Wikia.*[19] The most recent version, i.e., DBkWik version 1.1., integrates data extracted from 12,840 Wiki dumps, comprising 14,743,443 articles in total. While DBpedia relies on a manually created ontology and mappings to that, such an ontology does not exist for DBkWik, i.e., the schema for DBkWik needs to be inferred on the fly. Moreover, while for Wikipedia-centric knowledge graphs like DBpedia, there are rarely any duplicate entities (since each entity corresponds to exactly one page in Wikipedia), duplicates exist in DBkWik both on in the A-box and the T-box, and need to be handled in an additional integration step. Figure 2 illustrates the creation process of DBkWik.

---

[18] The ontology has the complexity *SRF(D)*, it mainly defines a class and a property hierarchy, together with domains, ranges, and disjointness statements.

[19] http://www.wikia.com/fandom.

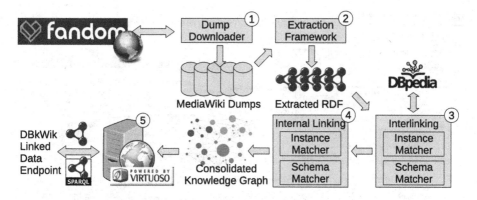

Fig. 2. The DBkWik creation process [26]

As a result, the latest release of DBkWik comprises 11M entities and 96M facts. With 12k classes and 129k relations, the schema is fairly detailed.

2.8 WebIsALOD

While the above knowledge graphs cover a multitude of different relations between entities, the WebIsALOD dataset is focusing solely on building a large-scale hierarchy of concepts, i.e., it builds a lattice of *hypernymy relations*. To that end, it scans the Common Crawl Web corpus[20] for so-called Hearst patterns, e.g. *X such as Y*, to infer relations like X skos:broader Y. The graph comes with very detailed provenance data, including the patterns used and the original text snippets, together with their sources [25]. Figure 3 depicts the schema of WebIsALOD, illustrating the richness of its metadata.

In total, WebIsALOD contains more than 212 million entities (however, it does not strictly separate between an *entity* and a *class*), and 400 million hypernymy relations.

Table 1 gives an overview of the knowledge graphs discussed above and their characteristics.[21] The table depicts the number of entities and relations, as well as the average indegree and outdegree of entity (i.e., the average number of ingoing and outgoing relations for each entity), as well as the size of the schema in terms of the number of classes and properties.

2.9 Non-public Knowledge Graphs

Furthermore, company-owned knowledge graphs exist, like the already mentioned Google Knowledge Graph, Google's Knowledge Vault [11], Yahoo's Knowledge Graph [2], Microsoft's Satori, and Facebook's Knowledge Graph. However, those are not publicly available, and hence neither suited to build applications by parties other than the owners, nor can they be analyzed in depth.

---

[20] http://commoncrawl.org/.
[21] The numbers are taken from [24,61].

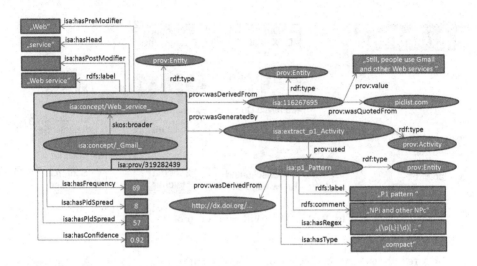

Fig. 3. The schema of the WebIsALOD knowledge graph [24]

Table 1. Public cross-domain knowledge graphs and their size

Knowledge graph	# Entities	# Facts	Avg. indegree	Avg. outdegree	# Classes	# Properties
OpenCyc	118,499	2,413,894	10.03	9.23	116,822	165
NELL	1,974,297	3,402,971	5.33	1.25	290	1,334
YAGO3	5,130,031	1,435,808,056	9.83	41.25	30,765	11,053
DBpedia	5,109,890	397,831,457	13.52	47.55	754	3,555
DBkWik	11,163,719	91,526,001	0.70	8.17	12,029	128,566
Wikidata	44,077,901	1,633,309,138	9.83	41.25	30,765	11,053
WebIsALOD	212,184,968	400,533,808	3.72	3.31	–	1

3 Using Machine Learning for Building and Refining Knowledge Graphs

As discussed above, large-scale knowledge graph can hardly be created manually [49]. Therefore, heuristics are quite frequently used in the creation of knowledge graphs, i.e., methods that can efficiently create large-scale knowledge graphs, trading off data volume for accuracy.

Machine learning methods can serve as a technique to implement such heuristics. They are widely used both in the creation of a knowledge graph (e.g., for NELL, whose creation is fully machine-learning based), as well as in the subsequent refinement of the generated knowledge graphs [47].

3.1   Type and Relation Prediction

No knowledge graph will ever contain every piece of knowledge that exists in the world. Therefore, all knowledge graphs, no matter whether they are created manually or heuristically, can have only *incomplete* knowledge.

Being grounded in semantic web standards, which adhere to the *open world assumption*, this is not a problem from a logical perspective. However, in many application scenarios, the value of a knowledge graph grows with the amount of knowledge encoded therein. Therefore, a lot of work as been devoted on *knowledge graph completion* [47].

As discussed above, semantic web knowledge graphs come with a common schema or ontology, which usually define a – sometimes deeper, sometimes more shallow – hierarchy of classes. Hence, adding missing type information is an important and frequently addressed task in knowledge graph completion.

One of the simplest and hence most scalable approaches is *SDType* [51]. SDType considers each relation in which an entity takes part as an indicator for its type. For example, the statement

```
Germany hasCapital Berlin.
```

involves the two entities `Germany` and `Berlin`, connected by the relation `hasCapital`. Each relation has a specific distribution of types of entities it connects. For example, considering all the pairs of entities connected by the relation `hasCapital` and analyzing their types, there is a high probability that the subject is of type `Country` and the object is of type `City`.

SDType computes a weighted average across all relations an entity (such as `Berlin`) is connected by, using their specific distribution of subject and object types. Here, most of the relations `Berlin` will have a high probability of the entity being a `City`. The weights are assigned by the specificity of the relation w.r.t. types. For example, the relation `hasCapital` is very specific for the subject and the object (it mostly links geographic regions to cities), whereas the relation `knownFor` is only specific for the subject (it is mainly used for subjects of type `Person`), but very unspecific for the object (persons can be known for lots of different things, i.e., the distribution of types for the object of `knownFor` is rather wide). This is illustrated in Fig. 4: when observing a `knownFor` relation between two entities, it is very likely that the subject is of type `Person`, whereas such a conclusion is more difficult for the object.

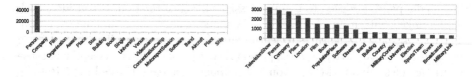

Fig. 4. Distribution of the subject (left) and object (right) types of the `knownFor` relation

Due to its simplicity, it is possible to develop well performing implementations of SDType, and the algorithm has meanwhile been integrated in DBpedia and used for building DBpedia releases.

The predictions of SDType are coherent with the existing type hierarchy, given that the type information is fully materialized on the input knowledge graph – i.e., it holds that for each subclass relation $R \sqsubseteq S$, if $R(a)$ is contained in the knowledge graph, then $S(a)$ is also contained. However, this is a characteristic by design, but SDType does not specifically exploit the hierarchy information.

Looking at type prediction from a machine learning perspective, it can be considered a hierarchical multi-label classification problem, i.e., a classification problem where each instance can belong to multiple classes, and those classes form a hierarchy [75]. In [36], we have shown that the type prediction problem can be well solved using a hierarchical classification problem. We used a *local classifier per node*, i.e., we train a classifier for each class in the hierarchy, sampling instances from the class' neighbors in the hierarchy as negative training examples. Since the problem is thereby decomposed into smaller learning problems, the solution becomes scalable and even largely parallelizable.

The scalability of type prediction using hierarchical classification can be even further improved when taking into account that only a small subset of features is required to tell a class from its siblings. For example, for telling a person from an organization, a few relations like `ceo`, `headquarter`, `birthplace`, and `nationality` are sufficient. For telling a movie from a book, relations like `director`, `studio`, `author`, and `publisher` are helpful. Incorporating local feature selection, i.e., determining a small subset of relevant features for each individual classification problem within the hierarchical classification, allows the approach to be scaled to very large knowledge graphs [34].

Type information is often used as a prediction target, but other relations are possible prediction targets as well. Usually, the knowledge graph itself is used as ground truth, i.e., every relation assertion in the knowledge graph is used as a positive training example. Since semantic web knowledge graphs follow the open world assumption, and machine learning classifiers usually expect both positive and negative examples, a common trick to generate negative examples is the so-called *partial completeness assumption* [15] or *local closed world assumption* [11]: it is assumed that if there is set of objects $o_1 \ldots o_n$ for a given subject s and a relation r, i.e., $r(s, o_1), r(s, o_2), \ldots r(o_n)$ are contained in the knowledge graph, then this information is complete w.r.t. s, i.e., there is no $o' \notin \{o_1, \ldots, o_n\}$ so that $r(s, o')$ holds.

Building on this assumption, the approach sketched in [23] uses abstracts in Wikipedia pages are used to train a classifier for each relation. It considers each entity linked within an abstract in a Wikipedia page as a candidate relation. Then, it uses a set of features to learn heuristic rules, such as: The first place to be mentioned in an article about a person is that person's birth place. Using those heuristic rules, about 1M additional statements could be learned for DBpedia. The approach has been shown to work for Wikipedia pages of any language, and even for other Wikis, and is therefore also applicable to DBkWik [22].

3.2 Error Detection

Most heuristics used for knowledge graph construction have to address a trade-off between size and accuracy of the resulting knowledge graph. Decent sized knowledge graphs can only be constructed heuristically by accepting a certain level of noise.

For that reason, a few approaches have been proposed to identify wrong facts in a knowledge graph. Again, a simplistic approach named *SDValidate* follows the same basic idea of SDType, i.e., using statistical distribution of types in the subject and object positions of a statement [52]. A statement is considered wrong by the approach if the types in the subject and object position differ significantly from the predicate's characteristic distribution.

While SDType is quite stable since it combines a lot of evidence (i.e., many relations in which an instance is involved) and improves with the connectivity of the underlying knowledge graph, SDValidate is less powerful since it usually has only fewer information to work with, i.e., the explicit types set for an entity are typically less than the average degree [61]. Therefore, additional evidence needs to be taken into account to reliably flag wrong relations.

PaTyBRED is a machine-learning based approach that does not only rely on type features, but also on paths in which an entity is involved. It uses relations in the knowledge graph as positive training examples, and creates negative examples by randomly replacing the subject or object with another instance from the knowledge graph.[22]

For training a model, *PaTyBRED* takes into account both types and paths, and therefore, it can also learn that the path $residence(X, Y), country(Y, Z)$ is positive evidence for the relation assertion $nationality(X, Z)$. We have shown that this combination outperforms both approaches based on types and based on paths alone.

An alternative to addressing error detection as a binary classification problem is assigning a confidence score to each entity. This approach is applied to building the WebIsALOD dataset. Here, we use a crowd-sourced gold standard of positive and negative examples (i.e., randomly selected examples from the initial extraction presented to human judges for validation) to train a classifier for telling correct from incorrect relations, using the rich provenance metadata. Instead of discarding the negative examples, we attach the classifier's confidence score as a confidence to each individual statement [24].

Both approaches – identifying and removing errors, as well as using confidence scores – have their advantages and disadvantages. Removing wrong statements leads to a clean and intuitively usable dataset, and also reduces that dataset's size, which can help in the processing. On the other hand, keeping all statements and attaching a confidence score has the advantage of letting the user set an individual threshold, thereby deciding on whether higher recall or

---

[22] While more complex strategies for generating meaningful negative training examples exist [29], we have observed no significant qualitative difference in the resulting models' accuracy to creating random negative examples, although those complex strategies are computationally much more expensive.

higher precision is required for a given task at hand. Furthermore, it opens the opportunity to process the dataset with methods being able to deal with such imprecision, e.g., probabilistic and possibilistic reasoners [55].

3.3 Approximate Local Reasoning

As discussed above, many knowledge graphs come with an ontology, which may be more or less formal. Given that a certain degree of formality is provided, i.e., the ontology does not only define a class hierarchy and domain and range restrictions for the properties, but also more restrictions, such as disjoint classes, the knowledge graph can be validated against the ontology. For example, if the two classes $City$ and $Team$ are defined as disjoint classes, the range of $playsFor$ is $Team$, and the two axioms $playsFor(Ronaldo, Madrid)$ and $City(Madrid)$ are defined in the knowledge graph, the combination of both axioms can be detected as a violation of the underlying ontology – although deciding $which$ of the two axioms is wrong is not possible with automatic reasoning.

Some knowledge graphs, such as DBpedia and NELL, validate new axioms against an existing ontology before adding them to the knowledge graph. For DBpedia, there also exists a mapping to the top level ontology $DOLCE$ [16], which can be used to detect more violations against the ontology. For example, the statement $award(TimBernersLee, RoyalSociety)$, together with $Organization$ $(RoyalSociety)$ and the range of $award$ being defined as $Award$, is not a violation against the DBpedia ontology, since there is no explicit disjointness between $Award$ and $Organization$. In contrast, the usage of DOLCE defines the corresponding super classes (i.e., $Description$ and $SocialAgent$) as disjoint. Hence, exploiting the additional formalism of the upper level ontology of DOLCE leads to detecting more inconsistencies [54].

An issue with using reasoning, or even local reasoning (i.e., reasoning only on a statement and its related statements) to detect inconsistencies, is computationally expensive: Validating all statements in DBpedia against the DBpedia and DOLCE ontology using a state of the art reasoner like HermiT [17] would take several weeks. In [57], we have introduced an approach that approximates reasoning-based fact checking by exploiting machine learning: we treat the validation problem as a binary classification problem, and let a reasoner validate a small number of statements as consistent or inconsistent. Those examples are then used to train a classifier, whose model approximates the reasoner's behavior. It has been shown that such models can reach an accuracy above 95%, at the same time being some orders of magnitude faster: instead of several weeks, the consistency of all statements in DBpedia can validated in less than two hours.

3.4 Deriving Higher Level Patterns

Once erroneous statements have been identified, there are several ways to proceed. For once, they can be simply removed from the knowledge graph. Second, as for WebIsALOD, they may get lower confidence ratings and defer the decision to a later point.

However, analyzing (potentially) erroneous statements more deeply can also reveal further insights. Clustering the errors may reveal groups of similar errors, which may have a common root [54]. Such common roots may be errors in the ontology, in particular translation statements (e.g., for DBpedia: mappings from a Wikipedia infobox to the common ontology) [60], or the handling of particular kinds of input, such as numerical values [14,80], dates, links with hashtags, etc. [48].

Once the root cause is identified, it is possible to track back the issue and address it at the respective place, i.e., improving the creation process of the knowledge graph. The advantage is that a group of errors can be addressed with a single fix, and the result is sustainable, i.e., it is also applied for any future version of the same knowledge graph [48]. Moreover, grouping errors can also serve as a sanity check: if an identified error does not belong to any group, i.e., it only occurs as a single, isolated statement, it is often a false negative, i.e., a wrongly identified error [54].

3.5 Evaluating Machine Learning on Knowledge Graphs

Although there is a larger body of work in applying machine learning methods to knowledge graph refinement, there is no common standard evaluation protocol and set of benchmarks. Methods encompass evaluation against (partial) gold standards, the knowledge graph itself (treated as a silver standard), and retrospective evaluations, i.e., evaluating the output of the knowledge graph. With respect to evaluation methods, precision and recall are quite frequently used, but other metrics, e.g., accuracy, area under the precision-recall curve (AUC-PR), area under the ROC curve (AUC-ROC), etc. [6], are also observed. As an additional dimension to result quality, computational performance can also be evaluated.

In [47], we have observed that a majority of all approaches is only evaluated against one knowledge graph, usually DBpedia (see Fig. 5). This often limits the significance of the results, because it is unclear whether the approach overfits and/or consciously or unconsciously exploits of the specific knowledge graph at hand. Therefore, evaluations against multiple knowledge graphs are clearly advised.

3.6 Partial Gold Standard

One common evaluation strategy is to use a partial gold standard. In this methodology, a subset of graph entities or relations are selected and labeled manually. Other evaluations use external knowledge graphs and/or databases as partial gold standards.

For completion tasks, this means that all axioms that *should exist in the knowledge graph* are collected, whereas for correction tasks, a set of axioms in the graph is manually labeled as correct or incorrect. The quality of completion approaches is usually measured in recall, precision, and F-measure, whereas for correction methods, accuracy and/or area under the ROC curve (AUC) are often used alternatively or in addition.

Sourcing partial gold standards from humans can lead to high quality data (given that the knowledge graph and the ontology it uses are not overly complex), but is costly, so that those gold standards are usually small. Exploiting other knowledge graphs based on knowledge graph interlinks (e.g., using Freebase data as a gold standard to evaluate DBpedia) is sometimes proposed to yield larger-scale gold standards, but has two sources of errors: errors in the target knowledge graph, and errors in the linkage between the two. For example, it has been reported that 20% of the interlinks between DBpedia and Freebase are incorrect [81], and that roughly half of the `owl:sameAs` links between knowledge graphs connect two things which are related, but not *exactly* the same (such as the company *Starbucks* and a particular Starbucks coffee shop) [19].

3.7   Knowledge Graph as Silver Standard

Another evaluation strategy is to use the given knowledge graph itself as a test dataset. Since the knowledge graph is not perfect (otherwise, refinement would not be necessary), it cannot be considered as a *gold standard*. However, assuming that the given knowledge graph is already of reasonable quality, we call this method *silver standard* evaluation, as already proposed in other works [18,27,45].

The silver standard method is usually applied to measure the performance of knowledge graph *completion* approaches, where it is analyzed how well relations in a knowledge graph can replicated by a knowledge graph completion method. As for gold standard evaluations, the result quality is usually measured in recall, precision, and F-measure. In contrast to using human annotations, large-scale evaluations are easily possible. The silver standard method is only suitable for evaluating knowledge graph completion, not for error detection, since it assumes the knowledge graph to be correct.

There are two variants of silver standard evaluations: in the more common ones, the entire knowledge graph is taken as input to the approach at hand, and the evaluation is then also carried out on the entire knowledge graph. As this may lead to an overfitting effect (in particular for internal methods), some works also foresee the splitting of the graph into a training and a test partition, which, however, is not as straight forward as, e.g., for propositional classification tasks [44], which is why most papers use the former method. Furthermore, split and cross validation do not fully solve the overfitting effect. For example, if a knowledge graph, by construction, has a bias towards certain kinds of information (e.g., relations are more complete for some classes than for others), approaches over adapting to that bias will be rated better than those which do not (and which may actually perform better in the general case).

Since the knowledge graph itself is not perfect, this evaluation method may sometimes underrate the evaluated approach. More precisely, most knowledge graphs follow the *open world assumption*, i.e., an axiom not present in the knowledge graph may or may not hold. Thus, if a completion approach correctly predicts the existence of an axiom missing in the knowledge graph, this would count as a false positive and thus lower precision. Approaches overfitting to the coverage bias of a knowledge graph at hand may thus be overrated.

3.8 Retrospective Evaluation

For retrospective evaluations, the output of a given approach is given to human judges for annotation, who then label suggested completions or identified errors as correct and incorrect. The quality metric is usually accuracy or precision, along with a statement about the total number of completions or errors found with the approach, and ideally also with a statement about the agreement of the human judges.

In many cases, automatic refinement methods lead to a very large number of findings, e.g., lists of tens of thousands of axioms which are potentially erroneous. Thus, retrospective evaluations are often carried out only on samples of the results. For some approaches which produce higher level artifacts – such as error patterns or completion rules – as intermediate results, a feasible alternative is to evaluate those artifacts instead of the actually affected axioms.

While partial gold standards can be reused for comparing different methods, this is not the case for retrospective evaluations. On the other hand, retrospective evaluations may make sense in cases where the interesting class is rare. For example, when evaluating error detection methods, a sample for a partial gold standard from a high-quality graph is likely not to contain a meaningful number of errors. In those cases, retrospective evaluation methodologies are often preferred over partial gold standards.

Another advantage of retrospective evaluations is that they allow a very detailed analysis of an approach's results. In particular, inspecting the errors made by an approach often reveals valuable findings about the advantages and limitations of a particular approach.

Table 2 sums up the different evaluation methodologies and contrasts their advantages and disadvantages.

Table 2. Overview on evaluation methods with their advantages and disadvantages [47]

Methodology	Advantages	Disadvantages
Partial gold standard	Highly reliable results reusable	Costly to produce balancing problems
Knowledge graph as silver standard	Large-scale evaluation feasible subjectiveness is minimized	Less reliable results prone to overfitting
Retrospective evaluation	Applicable to disbalanced problems allows for more detailed analysis of approaches	Not reusable approaches cannot be compared directly

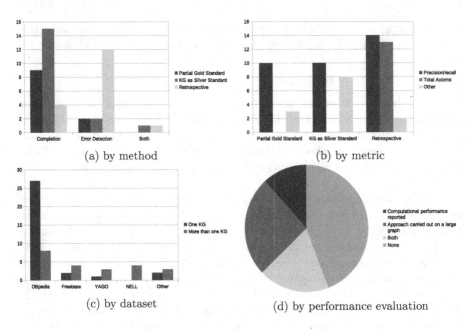

(a) by method   (b) by metric

(c) by dataset   (d) by performance evaluation

Fig. 5. Breakdown of evaluations observed in [47] by method, metrics, dataset, and computational performance evaluation

3.9 Computational Performance

In addition to the performance w.r.t. correctness and/or completeness of results, computational performance considerations become more important as knowledge graphs become larger. Typical performance measures for this aspect are runtime measurements, as well as memory consumption. Besides explicit measurement of computational performance, a "soft" indicator for computational performance is whether an approach has been evaluated (or at least the results have been materialized) on an entire large-scale knowledge graph, or only on a subgraph. The latter is often done when applying evaluations on a partial gold standard, where the retrospective approach is only executed on entities contained in that partial gold standard.

Furthermore, synthetic knowledge graphs, constructed with exactly specified characteristics, can assist in more systematic scalability testing [35].

4   Using Knowledge Graphs for Machine Learning

The common task of data mining is to discover patterns in data, which can be used either to gain a deeper understanding of the data (i.e., descriptive data mining), or for predictions of future developments (i.e., predictive data mining). For solving those tasks, machine learning methods are used.

Fayyad et al. have introduced a prototypical pipeline which leads from data to knowledge, comprising the steps of selection, preprocessing, transformation of data, applying some machine learning or data mining, and interpreting the results [13]. As depicted in Fig. 6 and discussed in detail in [69], Linked Data and, in particular, semantic web knowledge graphs can be used at almost every stage in the process: the first step is to link the data to be analyzed to the corresponding concepts in a knowledge graph. Once the local data is linked to a LOD dataset, we can explore the existing links in the knowledge graph pointing to related entities in the knowledge graph, as well as follow links to other graphs. In the next step, various techniques for data consolidation, preprocessing, and cleaning are applied, e.g., schema matching, data fusion, value normalization, treatment of missing values and outliers, etc. Next, some transformations on the collected data need to be performed in order to represent the data in a way that it can be processed with any arbitrary data analysis algorithms. Since most algorithms demand a *propositional* form of the input data (i.e., each entity being represented as a set of features), this usually includes a transformation of the knowledge *graph* to a canonical propositional form. After the data transformation is done, a suitable data mining algorithm is selected and applied on the data. In the final step, the results of the data mining process are presented to the user. Here, ease the interpretation and evaluation of the results of the data mining process, where knowledge graphs can be used as well [69].

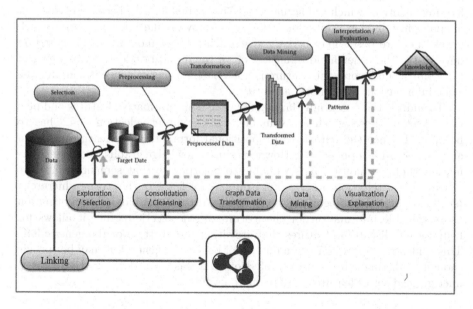

Fig. 6. Enhancing the data mining workflow by Fayyad et al. with Linked Data [69], based on [13]

4.1 Simple Feature Engineering

One of the main issues for exploiting knowledge graphs in machine learning tasks is that most machine learning algorithms are tailored towards propositional data, i.e., data where each instance is represented as a row in a table as a set of attribute values, also called *features*. In contrast, each instance in a knowledge graph is a node in a graph, connected to other nodes via edges. Hence, for being able to apply a machine learning algorithm to a set of instances in a knowledge graph, those have to be transferred into a propositional form first, a process which we call *propositionalization* [65].

Generally, there are two families of approaches for feature engineering from semantic web knowledge graphs: supervised and unsupervised [53].

For supervised approaches, knowledge about the knowledge graph at hand and its vocabulary is required. This can be either hard coded in the application, e.g., by adding features for genre and actors to a movie [7], or by letting the user specify queries to the knowledge graph, assuming that the user knows the graph's schema in depth [28].

Unsupervised approaches, on the other hand, do not make any assumptions about the dataset at hand. In contrast, they rather create features dynamically for all entities they encounter, e.g., by creating a numerical feature for each numerical literal, a set of binary features for each entity's classes, etc. These approaches have been shown to create valuable data mining features for a lot of machine learning problems, however, they may also lead to a very large set of features, many of which are actually not very valuable, e.g., because of sparsity or uniformity. For example, for a dataset of movies, the information on which novels they are based may be very *sparse* (since most movies are not based on novels). On the other hand, almost[23] of all them will be identified as being of type *Movie*. Therefore, this feature is not very informative, since it contains the same information for (almost) all entities.

To address the potentially large number of non-informative features and pick the subset of those which are valuable, feature subset selection [9,42] has to be applied. Once the data is in propositional form, standard feature selection algorithms can be applied [62]. However, since knowledge graphs also have information about semantics, it is valuable to incorporate that semantics into the feature subset selection process. For example, for features that form a hierarchy (such as types and categories), we have shown that this hierarchy information bears valuable information for the feature subset selection, since it allows for heuristically discarding features that are either too abstract or too generic [66]. This is illustrated in Fig. 7: for a dataset of persons in general, it may be specific enough to distinguish them by profession, whereas for a dataset of athletes, a finer grained set of features may be more useful.

4.2   Feature Vector Generation Using RDF2vec

As discussed above, simple feature creation techniques lead to a large number of features, which, at the same time, are often rather sparse, i.e., each feature in itself carries only little information. This lead us to the development of RDF2vec, a method for creating universal, re-usable dense feature vectors from knowledge

---

[23] Due to the open world assumption, it is likely that this does not hold for all movies.

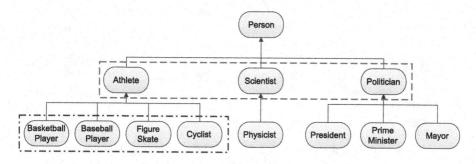

Fig. 7. A sample hierarchy of type features [66]

graphs. We have shown that even with moderately sized feature vectors (200 to 500 features), it is possible to outperform simple feature generation techniques on many tasks, at the same time allowing for re-using the same set of features across tasks, which minimizes the effort of feature generating for a new task [68, 70].

In a nutshell, RDF2vec picks up the idea of *word2vec* [39], which creates feature vector representations for word. Given a text corpus, word2vec trains a neural network for either predicting the surroundings of a word (context bag of words or CBOW variant), or a word, given its surroundings (skip-gram or SG variant). For example, for the sentence

Trent Reznor founded the band Nine Inch Nails in 1988

CBOW takes a single word (such as *Reznor*) as input and try to predict a set of probably surrounding words (such as *Trent, band, Nine, Inch*, etc.), while SG takes a set of words (e.g., *Trent, founded, the, band,...*) and tries to predict a word that "misses" in that set (e.g., *Reznor*).

The CBOW model predicts target words from context words within a given window. The model architecture is shown in Fig. 8a. The input layer is comprised from all the surrounding words for which the input vectors are retrieved from the input weight matrix, averaged, and projected in the projection layer. Then, using the weights from the output weight matrix, a score for each word in the vocabulary is computed, which is the probability of the word being a target word. Formally, given a sequence of training words $w_1, w_2, w_3, \ldots, w_T$, and a context window c, the objective of the CBOW model is to maximize the average log probability:

$$\frac{1}{T} \sum_{t=1}^{T} log\, p(w_t | w_{t-c} \ldots w_{t+c}), \tag{1}$$

where the probability $p(w_t | w_{t-c} \ldots w_{t+c})$ is calculated using the softmax function:

$$p(w_t | w_{t-c} \ldots w_{t+c}) = \frac{exp(\bar{v}^T v'_{w_t})}{\sum_{w=1}^{V} exp(\bar{v}^T v'_w)}, \tag{2}$$

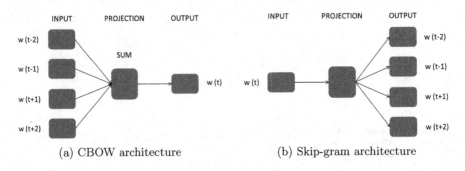

(a) CBOW architecture   (b) Skip-gram architecture

Fig. 8. Architecture of the CBOW and Skip-gram model [68]

where v'_w is the output vector of the word w, V is the complete vocabulary of words, and $\bar{v}$ is the averaged input vector of all the context words:

$$\bar{v} = \frac{1}{2c} \sum_{-c \leq j \leq c, j \neq 0} v_{w_{t+j}} \qquad (3)$$

The skip-gram model does the inverse of the CBOW model and tries to predict the context words from the target words (Fig. 8b). More formally, given a sequence of training words $w_1, w_2, w_3, \ldots, w_T$, and a context window c, the objective of the skip-gram model is to maximize the following average log probability:

$$\frac{1}{T} \sum_{t=1}^{T} \sum_{-c \leq j \leq c, j \neq 0} log p(w_{t+j}|w_t), \qquad (4)$$

where the probability $p(w_{t+j}|w_t)$ is calculated using the softmax function:

$$p(w_o|w_i) = \frac{exp(v'^T_{wo} v_{wi})}{\sum_{w=1}^{V} exp(v'^T_w v_{wi})}, \qquad (5)$$

where v_w and v'_w are the input and the output vector of the word w, and V is the complete vocabulary of words.

In both cases, calculating the softmax function is computationally inefficient, as the cost for computing is proportional to the size of the vocabulary. Therefore, two optimization techniques have been proposed, i.e., hierarchical softmax and negative sampling [40]. Empirical studies haven shown that in most cases negative sampling leads to a better performance than hierarchical softmax, which depends on the selected negative samples, but it has higher runtime.

Since knowledge graphs are not a text corpus, pseudo sentences are generated by performing random walks on the knowledge graph, and feeding the resulting sequences into the word2vec model.

Once the training is finished, all words (or, in our case, entities) are projected into a lower-dimensional feature space, and semantically similar words (or entities) are positioned close to each other.

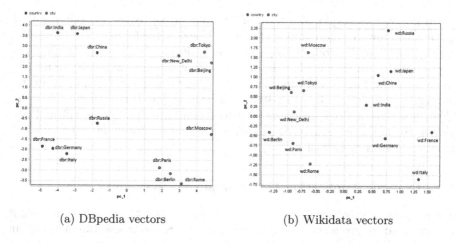

(a) DBpedia vectors   (b) Wikidata vectors

Fig. 9. Two-dimensional PCA projection of the 500-dimensional skip-gram vectors of countries and their capital cities [68]

The word2vec vector space representations for words have been shown to have two properties, among others: (1) semantically similar words are close in the resulting vector space, and (2) the direction of both grammatical as well as semantic relations between words remains stable for different pairs of words. Besides being able to use RDF2vec as a versatile and well performing feature vector generator, we have observed the same properties for RDF2vec as well, as depicted in Fig. 9.

To visualize the semantics of the vector representations, we employ Principal Component Analysis (PCA) to project the entities' feature vectors into a two dimensional feature space. We selected seven countries and their capital cities, and visualized their vectors as shown in Fig. 9. Figure 9a shows the corresponding DBpedia vectors, and Fig. 9b shows the corresponding Wikidata vectors. The figure illustrates the ability of the model to automatically organize entities of different types, and preserve the relationship between different entities. For example, we can see that there is a clear separation between the countries and the cities, and the relation *capital* between each pair of country and the corresponding capital city is preserved. Furthermore, we can observe that more similar entities are positioned closer to each other, e.g., we can see that the countries that are part of the EU are closer to each other, and the same applies for the Asian countries.

Using random walks for creating the sequences to be fed into the RDF2vec model is a straight forward and efficient approach, but has its pitfalls as well. Since not all entities and relations in a knowledge graph are equally important, putting more emphasis on the more important paths could lead to an improved embedding. However, telling an important from a less important relation, especially when attempting to create a task agnostic feature representation, is a hard problem.

In [8], we have explored a total of 12 different heuristics to guide the RDF graph walks towards more important paths, based on the frequency of properties, entities, and combinations of properties and entities, as well as the PageRank [3] of entities [76]. In that work, we could show that different heuristics, especially those based on PageRank, can improve the results in many cases, however, the interaction effects between the dataset or task at hand, the machine learning algorithm used, and the heuristic used to generate the paths are still underexplored, so that it is difficult to provide a decision guideline on which strategy works best for a given task.

4.3 Example Application 1: Recommender Systems

The purpose of recommender systems is to suggest items to users that they might be interested in, given the past interactions of users and items. Interactions can be implicit (e.g., users looking at items in an online store) or explicit (e.g., users buying items or rating them) [59].

Generally, there are two directions of recommender systems:

Collaborative filtering recommender systems, which rely on the interactions of users and items, and

Content based recommender systems, which rely on item similarity and suggest similar items.

Hybrid approaches, combining collaborative and content based mechanisms, exist as well. For collaborative filtering, there are two variants, i.e., user based and item based.

Content based approaches can benefit massively from knowledge graphs. Given that a dataset of items (e.g., movies, books, music titles) is linked to a knowledge graph, background information about those items can be retrieved from the knowledge graph, both in the form of direct interlinks between items (e.g., movies by the same director), as well as in the form of similar attributes (e.g., movies with a similar budget, runtime, etc.).

For the 2014 Linked Open Data-enabled recommender systems challenge [10], book recommendations had to be computed, with all the books in the dataset linked to DBpedia. There were two tasks: task 1 was the prediction of a user's rating given a user and a product, whereas task 2 was an actual recommendation task (i.e., proposing items to users). We made an experiment [63] using the software RapidMiner[24], together with two extensions: the Linked Open Data extension, which implements most of the algorithms discussed above [62], and the recommender system extension [38].

The features for content-based recommendation were extracted from DBpedia using the RapidMiner Linked Open Data extension. We use the following feature sets for describing a book:

---

[24] http://www.rapidminer.com/.

- All *direct types* of a book[25]
- All *categories of a book*
- All *categories of a book's author(s)*
- All *categories of a book including broader categories*[26]
- All *categories of a book's author(s) and of all other books* by the book's authors
- All *genres of a book* and of all other books by the book's authors
- All *authors that influenced or were influenced* by the book's authors
- A bag of words created from the *abstract* of the book in DBpedia. That bag of words is preprocessed by tokenization, stemming, removing tokens with less than three characters, and removing all tokens less frequent than 3% or more frequent than 80%.

Furthermore, we created a *combined book's feature set*, comprising direct types, qualified relations, genres and categories of the book itself, its previous and subsequent work and the author's notable work, the language and publisher, and the bag of words from the abstract.

Besides DBpedia, we made an effort to retrieve additional features from two additional LOD sources: British Library Bibliography (BLB) and DBTropes[27]. Using the RapidMiner LOD extension, we were able to link more than 90% of the books to BLB entities, but only 15% to DBTropes entities. However, the generated features from BLB were redundant with the features retrieved from DBpedia, and the coverage of DBTropes was too low to derive meaningful features. Hence, we did not pursue those sources further.

In addition to extracting content-based features, we used different generic recommenders in our approach. First, the RDF Book Mashup dataset[28] provides the average score assigned to a book on Amazon. Furthermore, DBpedia provides the number of ingoing links to the Wikipedia article corresponding to a DBpedia instance, and the number of links to other datasets (e.g., other language editions of DBpedia), which we also use as global popularity measures. Finally, SubjectiveEye3D delivers a subjective importance score computed from Wikipedia usage information[29].

To combine all feature sets into a content-based recommender engine, we trained recommender systems on each of the feature sets individually. In order to create a more sophisticated combination of the different base and generic recommenders, we trained a *stacking* model as described in [77]: We trained the base recommenders in 10 rounds in a cross validation like setting, collected their predictions, and learned a stacking model on the predictions. As stacking

---

[25] This includes types in the YAGO ontology, which can be quite specific (e.g., *American Thriller Novels*).

[26] The reason for not including broader categories by default is that the category graph is not a cycle-free tree, with some subsumptions being rather questionable.

[27] http://bnb.data.bl.uk/ and http://skipforward.opendfki.de/wiki/DBTropes.

[28] http://wifo5-03.informatik.uni-mannheim.de/bizer/bookmashup/.

[29] https://github.com/paulhoule/telepath/wiki/SubjectiveEye3D.

Table 3. Performances of the base and generic recommenders, the number of features used for each base recommender, and the performance of the combined recommenders

Recommender	#Features	Task 1		Task 2
		RMSE	LR β	F-Score
Item-based collaborative filtering	–	0.8843	+0.269	0.5621
User-based collaborative filtering	–	0.9475	+0.145	0.5483
Book's direct types	534	0.8895	−0.230	0.5583
Author's categories	2,270	0.9183	+0.098	0.5576
Book's (and author's author books') genres	582	0.9198	+0.082	0.5567
Combined book's properties	4,372	0.9421	+0.0196	0.5557
Author and influenced/influencedBy authors	1,878	0.9294	+0.122	0.5534
Books' categories and broader categories	1,987	0.939	+0.012	0.5509
Abstract bag of words	227	0.8893	+0.124	0.5609
RDT recommender on combined book's properties	4,372	0.9223	+0.128	0.5119
Amazon rating	–	1.037	+0.155	0.5442
Ingoing Wikipedia links	–	3.9629	+0.001	0.5377
SubjectiveEye3D score	–	3.7088	+0.001	0.5369
Links to other datasets	–	3.3211	+0.001	0.5321
Average of all individual recommenders	14	0.8824	–	–
Stacking with linear regression	14	0.8636	–	0.4645
Stacking with RDT	14	**0.8632**	–	0.4966
Borda rank aggregation	14	–	–	**0.5715**

models, we use linear regression and random decision trees [82], a variant of random forests, and for rank aggregation, we also use Borda rank aggregation. Table 3 shows the results, showing both the RMSE and F1 score on the two tasks of the challenge, as well as the weight β computed for each feature by the linear regression meta learner.

The main learnings from the experiment were:

- Knowledge graphs can provide additional background knowledge, and content based recommender systems trained using that background knowledge outperform collaborative filtering approaches.
- Combining predictions with a reasonably sophisticated stacking method outperforms simple averaging.
- Generic recommenders (i.e., global item popularity ranks) that are not taking into account the user's preferences are a surprisingly strong baseline.

In [71], we have explored the use of RDF2vec vector models as a means for feature generation for recommender systems. We have shown that content-based methods based on RDF2vec do not only outperform other feature generation techniques, but also state of the art collaborative filtering recommender systems.

Accordingly to Fig. 9a, Fig. 10 depicts a 2-dimensional PCA projection of a set of movies in the recommendation dataset. We can observe that related movies (Disney movies, Star Trek movies, etc.) have a low distance in the vector space, which allows for the use of content-based recommender systems based on item similarity in the vector space.

4.4 Example Application 2: Explaining Statistics

While recommender systems are an example for predicting machine learning, i.e., training a model for predicting future behavior of users, semantic web knowledge graphs can also be utilized in descriptive machine learning, where the task is to understand a dataset. One example for this is the interpretation of a given dataset, e.g., a collection of statistical observations.

One of the first prototypes for explaining statistical data with semantic web knowledge graphs was *Explain-a-LOD* [46]. The tool takes as input a statistics file, consisting of entities (e.g., regions) and a target variable (e.g., unemployment). It then uses a knowledge graph like DBpedia to identify factors that can be used to explain the target variable, e.g., by measuring correlation.

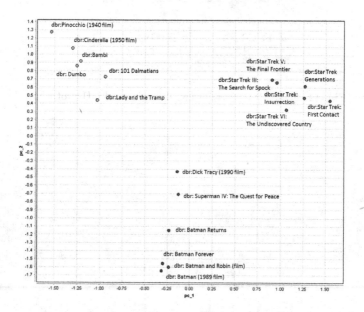

Fig. 10. PCA projection of example movies

As an example, we used a dataset of unemployment in different regions in France. Among others, Explain-a-LOD found the following (positively and negatively) correlating factors, using DBpedia as a starting point, and following links to other knowledge graphs and linked open datasets [64] (Fig. 11):

- gross domestic product (GDP) (negative)
- available household income (negative)
- R&D spendings (negative)
- energy consumption (negative)

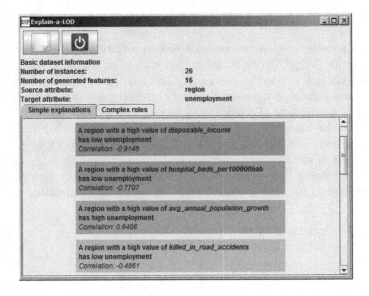

Fig. 11. Screenshot of the original Explain-a-LOD prototype

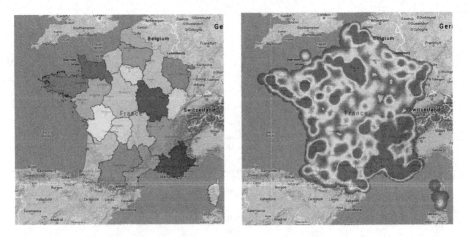

Fig. 12. Visualization of unemployment (left) and density of police stations (right)

- population growth (positive)
- casualties in road accidents (negative)
- number of fast food restaurants (positive)
- number of police stations (positive).

In order to further inspect the findings and allowing the user to more intuitively interact with such interpretations, it is also possible to graphically visualize the correlations, as shown in Fig. 12 [67].

While embedding based models have been shown to work well on predictive tasks, their usage in descriptive scenarios is rather limited in contrast. Even if we could find that certain attributes generated by an embedding model explain a statistical variable well, this would not be very descriptive, since the embedding dimensions do not come with a semantic annotation, and hence are not interpretable by humans.

5   Conclusion and Future Directions

In this chapter, we have looked at a possible symbiosis of machine learning and semantic web knowledge graphs from two angles: (1) using machine learning for knowledge graphs, and (2) using knowledge graphs in machine learning. In both areas, a larger body of works exist, and both are active and vital areas of research.

Our recent approaches to creating knowledge graphs which are complementary to those based on Wikipedia, i.e., WebIsALOD and DBkWik, have shown that there is an interesting potential of generating such knowledge graphs. Whenever creating and extending a knowledge graph or mapping a knowledge graph to existing ones, machine learning techniques can be used, either by having a training set manually curated, or by using knowledge already present in the knowledge graph for training models that add new knowledge or validate the existing information.

With respect to knowledge graph creation, there are still valuable sources on the Web that can be used. The magnitude of Wikis, as utilized by DBkWik, is just one direction, while there are other sources, like structured annotations on Web pages [37] or web tables [31], which can be utilized [78]. Also for Wiki-based extractions, not all information is used to date, e.g., with the potential of tables, lists, and enumerations still being underexplored [43,56].

While knowledge graphs are often developed in isolation, an interesting approach would be to use them as training data to improve each other, allowing cross fertilization of knowledge graphs. For example, WebIsALOD requires training data for telling instances and categories from each other, which could be gathered from DBpedia or YAGO. On the other hand, the latter are often incomplete with type information, which could be mined using features from WebIsALOD.

As discussed above, embedding methods are currently not usable for descriptive machine learning. Closing the gap between embeddings (which produce highly valuable features) and simple, but semantics preserving feature generation methods would help developing a new breed of descriptive machine learning methods [50]. To that end, embeddings either need to be semantically enriched with a posteriori developments, or trained in a fashion (e.g., by using pattern or rule learners, such as [21]) that allow for creating embedding spaces which are semantically meaningful.

Acknowledgements. I would like to thank (in alphabetical order) Aldo Gangemi, André Melo, Christian Bizer, Daniel Ringler, Eneldo Loza Mencía, Heiner Stuckenschmidt, Jessica Rosati, Johanna Völker, Julian Seitner, Kai Eckert, Michael Cochez, Nicolas Heist, Petar Ristoski, Renato De Leone, Robert Meusel, Simone Paolo Ponzetto, Stefano Faralli, Sven Hertling, and Tommaso Di Noia for their valuable input to this paper.

References

1. Bizer, C., Heath, T., Berners-Lee, T.: Linked data-the story so far. Int. J. Semant. Web Inf. Syst. **5**(3), 1–22 (2009)
2. Blanco, R., Cambazoglu, B.B., Mika, P., Torzec, N.: Entity recommendations in web search. In: Alani, H. (ed.) ISWC 2013. LNCS, vol. 8219, pp. 33–48. Springer, Heidelberg (2013). https://doi.org/10.1007/978-3-642-41338-4_3
3. Brin, S., Page, L.: The anatomy of a large-scale hypertextual web search engine. Comput. Netw. ISDN Syst. **30**(1), 107–117 (1998)
4. Bryl, V., Bizer, C.: Learning conflict resolution strategies for cross-language Wikipedia data fusion. In: Proceedings of the 23rd International Conference on World Wide Web, pp. 1129–1134. ACM (2014)
5. Carlson, A., Betteridge, J., Wang, R.C., Hruschka Jr., E.R., Mitchell, T.M.: Coupled semi-supervised learning for information extraction. In: Proceedings of the Third ACM International Conference on Web Search and Data Mining, pp. 101–110 (2010)
6. Caruana, R., Niculescu-Mizil, A.: Data mining in metric space: an empirical analysis of supervised learning performance criteria. In: Proceedings of the Tenth ACM SIGKDD International Conference on Knowledge Discovery and Data Mining, pp. 69–78. ACM (2004)
7. Cheng, W., Kasneci, G., Graepel, T., Stern, D., Herbrich, R.: Automated feature generation from structured knowledge. In: 20th ACM Conference on Information and Knowledge Management, CIKM 2011 (2011)
8. Cochez, M., Ristoski, P., Ponzetto, S.P., Paulheim, H.: Biased graph walks for RDF graph embeddings. In: Proceedings of the 7th International Conference on Web Intelligence, Mining and Semantics, p. 21. ACM (2017)
9. Dash, M., Liu, H.: Feature selection for classification. Intell. Data Anal. **1**(3), 131–156 (1997)
10. Di Noia, T., Cantador, I., Ostuni, V.C.: Linked open data-enabled recommender systems: ESWC 2014 challenge on book recommendation. In: Presutti, V. (ed.) SemWebEval 2014. CCIS, vol. 475, pp. 129–143. Springer, Cham (2014). https://doi.org/10.1007/978-3-319-12024-9_17
11. Dong, X., et al.: Knowledge vault: a web-scale approach to probabilistic knowledge fusion. In: Proceedings of the 20th ACM SIGKDD International Conference on Knowledge Discovery and Data Mining, pp. 601–610. ACM (2014)
12. Ehrlinger, L., Wöß, W.: Towards a definition of knowledge graphs. In: SEMANTiCS (2016)

13. Fayyad, U.M., Piatetsky-Shapiro, G., Smyth, P.: Advances in Knowledge Discovery and Data Mining, pp. 1–34. American Association for Artificial Intelligence, Menlo Park (1996). http://dl.acm.org/citation.cfm?id=257938.257942
14. Fleischhacker, D., Paulheim, H., Bryl, V., Völker, J., Bizer, C.: Detecting errors in numerical linked data using cross-checked outlier detection. In: Mika, P. (ed.) ISWC 2014. LNCS, vol. 8796, pp. 357–372. Springer, Cham (2014). https://doi.org/10.1007/978-3-319-11964-9_23
15. Galárraga, L.A., Teflioudi, C., Hose, K., Suchanek, F.: AMIE: association rule mining under incomplete evidence in ontological knowledge bases. In: 22nd International Conference on World Wide Web, pp. 413–422 (2013)
16. Gangemi, A., Guarino, N., Masolo, C., Oltramari, A., Schneider, L.: Sweetening ontologies with DOLCE. In: Gómez-Pérez, A., Benjamins, V.R. (eds.) EKAW 2002. LNCS, vol. 2473, pp. 166–181. Springer, Heidelberg (2002). https://doi.org/10.1007/3-540-45810-7_18
17. Glimm, B., Horrocks, I., Motik, B., Stoilos, G., Wang, Z.: HermiT: an OWL 2 reasoner. J. Autom. Reason. **53**(3), 245–269 (2014)
18. Groza, T., Oellrich, A., Collier, N.: Using silver and semi-gold standard corpora to compare open named entity recognisers. In: IEEE International Conference on Bioinformatics and Biomedicine, BIBM, pp. 481–485. IEEE, Piscataway (2013). https://doi.org/10.1109/BIBM.2013.6732541
19. Halpin, H., Hayes, P.J., McCusker, J.P., McGuinness, D.L., Thompson, H.S.: When owl:sameAs isn't the same: an analysis of identity in linked data. In: Patel-Schneider, P.F. (ed.) ISWC 2010. LNCS, vol. 6496, pp. 305–320. Springer, Heidelberg (2010). https://doi.org/10.1007/978-3-642-17746-0_20
20. Hearst, M.A.: Automatic acquisition of hyponyms from large text corpora. In: Proceedings of the 14th Conference on Computational Linguistics, vol. 2, pp. 539–545. Association for Computational Linguistics (1992)
21. Hees, J., Bauer, R., Folz, J., Borth, D., Dengel, A.: An evolutionary algorithm to learn SPARQL queries for source-target-pairs: finding patterns for human associations in DBpedia. CoRR abs/1607.07249 (2016). http://arxiv.org/abs/1607.07249
22. Heist, N., Hertling, S., Paulheim, H.: Language-agnostic relation extraction from abstracts in Wikis. Information **9**(4), 75 (2018)
23. Heist, N., Paulheim, H.: Language-agnostic relation extraction from Wikipedia abstracts. In: d'Amato, C. (ed.) ISWC 2017. LNCS, vol. 10587, pp. 383–399. Springer, Cham (2017). https://doi.org/10.1007/978-3-319-68288-4_23
24. Hertling, S., Paulheim, H.: WebIsALOD: providing hypernymy relations extracted from the web as linked open data. In: d'Amato, C. (ed.) ISWC 2017. LNCS, vol. 10588, pp. 111–119. Springer, Cham (2017). https://doi.org/10.1007/978-3-319-68204-4_11
25. Hertling, S., Paulheim, H.: Provisioning and usage of provenance data in the WEBIsALOD knowledge graph. In: First International Workshop on Contextualized Knowledge Graphs (2018)
26. Hofmann, A., Perchani, S., Portisch, J., Hertling, S., Paulheim, H.: DBkWik: towards knowledge graph creation from thousands of Wikis. In: International Semantic Web Conference (Posters and Demos) (2017)
27. Kang, N., van Mulligen, E.M., Kors, J.A.: Training text chunkers on a silver standard corpus: can silver replace gold? BMC Bioinform. **13**(1), 17 (2012). https://doi.org/10.1186/1471-2105-13-17
28. Narasimha, V., Kappara, P., Ichise, R., Vyas, O.P.: LiDDM: a data mining system for linked data. In: Workshop on Linked Data on the Web, LDOW 2011 (2011)

29. Lao, N., Cohen, W.W.: Relational retrieval using a combination of path-constrained random walks. Mach. Learn. **81**(1), 53–67 (2010)
30. Lehmann, J., et al.: DBpedia - a large-scale, multilingual knowledge base extracted from Wikipedia. Semant. Web J. **6**(2), 167–195 (2013)
31. Lehmberg, O., Ritze, D., Meusel, R., Bizer, C.: A large public corpus of web tables containing time and context metadata. In: Proceedings of the 25th International Conference Companion on World Wide Web, pp. 75–76. International World Wide Web Conferences Steering Committee (2016)
32. Lenat, D.B.: CYC: a large-scale investment in knowledge infrastructure. Commun. ACM **38**(11), 33–38 (1995)
33. Mahdisoltani, F., Biega, J., Suchanek, F.M.: YAGO3: a knowledge base from multilingual Wikipedias. In: CIDR (2013)
34. Melo, A., Paulheim, H.: Local and global feature selection for multilabel classification with binary relevance. Artif. Intell. Rev. 1–28 (2017)
35. Melo, A., Paulheim, H.: Synthesizing knowledge graphs for link and type prediction benchmarking. In: Blomqvist, E., Maynard, D., Gangemi, A., Hoekstra, R., Hitzler, P., Hartig, O. (eds.) ESWC 2017. LNCS, vol. 10249, pp. 136–151. Springer, Cham (2017). https://doi.org/10.1007/978-3-319-58068-5_9
36. Melo, A., Paulheim, H., Völker, J.: Type prediction in RDF knowledge bases using hierarchical multilabel classification. In: Proceedings of the 6th International Conference on Web Intelligence, Mining and Semantics, p. 14. ACM (2016)
37. Meusel, R., Petrovski, P., Bizer, C.: The WebDataCommons microdata, RDFa and microformat dataset series. In: Mika, P. (ed.) ISWC 2014. LNCS, vol. 8796, pp. 277–292. Springer, Cham (2014). https://doi.org/10.1007/978-3-319-11964-9_18
38. Mihelčić, M., Antulov-Fantulin, N., Bošnjak, M., Šmuc, T.: Extending RapidMiner with recommender systems algorithms. In: RapidMiner Community Meeting and Conference, RCOMM 2012 (2012)
39. Mikolov, T., Chen, K., Corrado, G., Dean, J.: Efficient estimation of word representations in vector space. arXiv preprint arXiv:1301.3781 (2013)
40. Mikolov, T., Sutskever, I., Chen, K., Corrado, G.S., Dean, J.: Distributed representations of words and phrases and their compositionality. In: Advances in Neural Information Processing Systems, pp. 3111–3119 (2013)
41. Miller, G.A.: WordNet: a lexical database for English. Commun. ACM **38**(11), 39–41 (1995). https://doi.org/10.1145/219717.219748
42. Molina, L.C., Belanche, L., Nebot, À.: Feature selection algorithms: a survey and experimental evaluation. In: International Conference on Data Mining, ICDM, pp. 306–313. IEEE (2002)
43. Muñoz, E., Hogan, A., Mileo, A.: Using linked data to mine RDF from Wikipedia's tables. In: Proceedings of the 7th ACM International Conference on Web Search and Data Mining, pp. 533–542. ACM (2014)
44. Neville, J., Jensen, D.: Iterative classification in relational data. In: Proceedings of the AAAI-2000 Workshop on Learning Statistical Models from Relational Data, pp. 13–20. AAAI, Palo Alto (2000). http://www.aaai.org/Library/Workshops/2000/ws00-06-007.php
45. Nothman, J., Ringland, N., Radford, W., Murphy, T., Curran, J.R.: Learning multilingual named entity recognition from Wikipedia. Artif. Intell. **194**, 151–175 (2013). https://doi.org/10.1016/j.artint.2012.03.006
46. Paulheim, H.: Generating possible interpretations for statistics from linked open data. In: Simperl, E., Cimiano, P., Polleres, A., Corcho, O., Presutti, V. (eds.) ESWC 2012. LNCS, vol. 7295, pp. 560–574. Springer, Heidelberg (2012). https://doi.org/10.1007/978-3-642-30284-8_44

47. Paulheim, H.: Knowledge graph refinement: a survey of approaches and evaluation methods. Semant. Web **8**(3), 489–508 (2017)
48. Paulheim, H.: Data-driven joint debugging of the DBpedia mappings and ontology. In: Blomqvist, E., Maynard, D., Gangemi, A., Hoekstra, R., Hitzler, P., Hartig, O. (eds.) ESWC 2017. LNCS, vol. 10249, pp. 404–418. Springer, Cham (2017). https://doi.org/10.1007/978-3-319-58068-5_25
49. Paulheim, H.: How much is a triple? - estimating the cost of knowledge graph creation. In: ISWC Blue Sky Ideas (2018, to appear)
50. Paulheim, H.: Make embeddings semantic again! In: ISWC Blue Sky Ideas (2018, to appear)
51. Paulheim, H., Bizer, C.: Type inference on noisy RDF data. In: Alani, H., et al. (eds.) ISWC 2013. LNCS, vol. 8218, pp. 510–525. Springer, Heidelberg (2013). https://doi.org/10.1007/978-3-642-41335-3_32
52. Paulheim, H., Bizer, C.: Improving the quality of linked data using statistical distributions. Int. J. Semant. Web Inf. Syst. (IJSWIS) **10**(2), 63–86 (2014)
53. Paulheim, H., Fürnkranz, J.: Unsupervised generation of data mining features from linked open data. In: International Conference on Web Intelligence, Mining, and Semantics, WIMS 2012 (2012)
54. Paulheim, H., Gangemi, A.: Serving DBpedia with DOLCE – more than just adding a cherry on top. In: Arenas, M. (ed.) ISWC 2015. LNCS, vol. 9366, pp. 180–196. Springer, Cham (2015). https://doi.org/10.1007/978-3-319-25007-6_11
55. Paulheim, H., Pan, J.Z.: Why the semantic web should become more imprecise (2012)
56. Paulheim, H., Ponzetto, S.P.: Extending DBpedia with Wikipedia list pages. In: NLP-DBPEDIA ISWC 2013 (2013)
57. Paulheim, H., Stuckenschmidt, H.: Fast approximate A-box consistency checking using machine learning. In: Sack, H., Blomqvist, E., d'Aquin, M., Ghidini, C., Ponzetto, S.P., Lange, C. (eds.) ESWC 2016. LNCS, vol. 9678, pp. 135–150. Springer, Cham (2016). https://doi.org/10.1007/978-3-319-34129-3_9
58. Pellissier Tanon, T., Vrandečić, D., Schaffert, S., Steiner, T., Pintscher, L.: From freebase to Wikidata: the great migration. In: Proceedings of the 25th International Conference on World Wide Web, pp. 1419–1428 (2016)
59. Ricci, F., Rokach, L., Shapira, B.: Introduction to recommender systems handbook. In: Ricci, F., Rokach, L., Shapira, B., Kantor, P.B. (eds.) Recommender Systems Handbook, pp. 1–35. Springer, Boston, MA (2011). https://doi.org/10.1007/978-0-387-85820-3_1
60. Rico, M., Mihindukulasooriya, N., Kontokostas, D., Paulheim, H., Hellmann, S., Gómez-Pérez, A.: Predicting incorrect mappings: a data-driven approach applied to DBpedia (2018)
61. Ringler, D., Paulheim, H.: One knowledge graph to rule them all? Analyzing the differences between DBpedia, YAGO, Wikidata & co. In: Kern-Isberner, G., Fürnkranz, J., Thimm, M. (eds.) KI 2017. LNCS, vol. 10505, pp. 366–372. Springer, Cham (2017). https://doi.org/10.1007/978-3-319-67190-1_33
62. Ristoski, P., Bizer, C., Paulheim, H.: Mining the web of linked data with Rapid-Miner. Web Semant.: Sci. Serv. Agents World Wide Web **35**, 142–151 (2015)
63. Ristoski, P., Loza Mencía, E., Paulheim, H.: A hybrid multi-strategy recommender system using linked open data. In: Presutti, V. (ed.) SemWebEval 2014. CCIS, vol. 475, pp. 150–156. Springer, Cham (2014). https://doi.org/10.1007/978-3-319-12024-9_19
64. Ristoski, P., Paulheim, H.: Analyzing statistics with background knowledge from linked open data. In: Workshop on Semantic Statistics (2013)

65. Ristoski, P., Paulheim, H.: A comparison of propositionalization strategies for creating features from linked open data. In: Linked Data for Knowledge Discovery, p. 6 (2014)
66. Ristoski, P., Paulheim, H.: Feature selection in hierarchical feature spaces. In: Džeroski, S., Panov, P., Kocev, D., Todorovski, L. (eds.) DS 2014. LNCS, vol. 8777, pp. 288–300. Springer, Cham (2014). https://doi.org/10.1007/978-3-319-11812-3_25
67. Ristoski, P., Paulheim, H.: Visual analysis of statistical data on maps using linked open data. In: Gandon, F., Guéret, C., Villata, S., Breslin, J., Faron-Zucker, C., Zimmermann, A. (eds.) ESWC 2015. LNCS, vol. 9341, pp. 138–143. Springer, Cham (2015). https://doi.org/10.1007/978-3-319-25639-9_27
68. Ristoski, P., Paulheim, H.: RDF2Vec: RDF graph embeddings for data mining. In: Groth, P. (ed.) ISWC 2016. LNCS, vol. 9981, pp. 498–514. Springer, Cham (2016). https://doi.org/10.1007/978-3-319-46523-4_30
69. Ristoski, P., Paulheim, H.: Semantic web in data mining and knowledge discovery: a comprehensive survey. Web Semant.: Sci. Serv. Agents World Wide Web **36**, 1–22 (2016)
70. Ristoski, P., Rosati, J., Noia, T.D., Leone, R.D., Paulheim, H.: RDF2Vec: RDF graph embeddings and their applications. Semant. Web (2018)
71. Rosati, J., Ristoski, P., Di Noia, T., Leone, R.D., Paulheim, H.: RDF graph embeddings for content-based recommender systems. In: CEUR Workshop Proceedings, vol. 1673, pp. 23–30. RWTH (2016)
72. Sarjant, S., Legg, C., Robinson, M., Medelyan, O.: "All you can eat" ontology-building: feeding Wikipedia to Cyc. In: Proceedings of the 2009 IEEE/WIC/ACM International Joint Conference on Web Intelligence and Intelligent Agent Technology, vol. 01, pp. 341–348. IEEE Computer Society, Piscataway (2009). https://doi.org/10.1109/WI-IAT.2009.60
73. Schmachtenberg, M., Bizer, C., Paulheim, H.: Adoption of the linked data best practices in different topical domains. In: Mika, P. (ed.) ISWC 2014. LNCS, vol. 8796, pp. 245–260. Springer, Cham (2014). https://doi.org/10.1007/978-3-319-11964-9_16
74. Seitner, J., et al.: A large database of hypernymy relations extracted from the web. In: LREC (2016)
75. Silla Jr., C.N., Freitas, A.A.: A survey of hierarchical classification across different application domains. Data Min. Knowl. Discov. **22**(1–2), 31–72 (2011)
76. Thalhammer, A., Rettinger, A.: PageRank on Wikipedia: towards general importance scores for entities. In: Sack, H., Rizzo, G., Steinmetz, N., Mladenić, D., Auer, S., Lange, C. (eds.) ESWC 2016. LNCS, vol. 9989, pp. 227–240. Springer, Cham (2016). https://doi.org/10.1007/978-3-319-47602-5_41
77. Ting, K.M., Witten, I.H.: Issues in stacked generalization. Artif. Intell. Res. **10**(1), 271–289 (1999)
78. Tonon, A., Felder, V., Difallah, D.E., Cudré-Mauroux, P.: VoldemortKG: mapping schema.org and web entities to linked open data. In: Groth, P. (ed.) ISWC 2016. LNCS, vol. 9982, pp. 220–228. Springer, Cham (2016). https://doi.org/10.1007/978-3-319-46547-0_23
79. Vrandečić, D., Krötzsch, M.: Wikidata: a free collaborative knowledge base. Commun. ACM **57**(10), 78–85 (2014)
80. Wienand, D., Paulheim, H.: Detecting incorrect numerical data in DBpedia. In: Presutti, V., d'Amato, C., Gandon, F., d'Aquin, M., Staab, S., Tordai, A. (eds.) ESWC 2014. LNCS, vol. 8465, pp. 504–518. Springer, Cham (2014). https://doi.org/10.1007/978-3-319-07443-6_34

81. Zaveri, A., et al.: User-driven quality evaluation of DBpedia. In: 9th International Conference on Semantic Systems, I-SEMANTICS 2013, pp. 97–104. ACM, New York (2013). https://doi.org/10.1145/2506182.2506195
82. Zhang, X., Yuan, Q., Zhao, S., Fan, W., Zheng, W., Wang, Z.: Multi-label classification without the multi-label cost. In: Proceedings of the Tenth SIAM International Conference on Data Mining (2010)
83. Zimmermann, A., Gravier, C., Subercaze, J., Cruzille, Q.: Nell2RDF: read the web, and turn it into RDF. In: Knowledge Discovery and Data Mining Meets Linked Open Data. CEUR Workshop Proceedings, vol. 992, pp. 2–8 (2013). http://ceur-ws.org/Vol-992/

Rule Induction and Reasoning
over Knowledge Graphs

Daria Stepanova$^{(\boxtimes)}$, Mohamed H. Gad-Elrab, and Vinh Thinh Ho

Max Planck Institute for Informatics, Saarland Informatics Campus,
Saarbrücken, Germany
{dstepano,gadelrab,hvthinh}@mpi-inf.mpg.de

Abstract. Advances in information extraction have enabled the automatic construction of large knowledge graphs (KGs) like DBpedia, Freebase, YAGO and Wikidata. Learning rules from KGs is a crucial task for KG completion, cleaning and curation. This tutorial presents state-of-the-art rule induction methods, recent advances, research opportunities as well as open challenges along this avenue. We put a particular emphasis on the problems of learning exception-enriched rules from highly biased and incomplete data. Finally, we discuss possible extensions of classical rule induction techniques to account for unstructured resources (e.g., text) along with the structured ones.

1 Introduction

Motivation. Recent advances in information extraction have led to huge graph-structured knowledge bases (KBs) also known as knowledge graphs (KGs) such as NELL [47], DBpedia [42], YAGO [72] and Wikidata [77]. These KGs contain millions or billions of relational facts in the form of subject-predicate-object (SPO) triples, e.g., $\langle$*albert_einstein marriedTo mileva_maric*$\rangle$ or $\langle$*albert_einstein type phycisist*$\rangle$. Such triples can be straightforwardly represented by means of positive unary and binary first-order logic facts, e.g. *marriedTo*(*albert_einstein,mileva_maric*) and *phycisist*(*albert_einstein*).

An important task over KGs is rule learning, which is relevant for a variety of applications ranging from knowledge graph curation (completion, error detection) [55] to data mining and semantic culturomics [73]. Rules over KGs are of the form *head* ← *body*, where *head* is a binary atom and *body* is a conjunction of (possibly negated) binary and unary atoms.

Traditionally, rule induction has been studied in the context of relational data mining in databases (see e.g., [58] for overview), and has recently been adapted to KGs (e.g., [6,30,79]). The methods from this area can be used to identify prominent patterns from KGs, such as *"Married people live in the same place"*, and cast them in the form of Horn rules (i.e., rules with only positive atoms), such as: $r_1 : livesIn(Y, Z) \leftarrow marriedTo(X, Y), livesIn(X, Z)$.

For the KG curation, this has two-fold benefits. First, since KGs operate under the open world assumption (i.e., absent facts are treated as unknown

© Springer Nature Switzerland AG 2018
C. d'Amato and M. Theobald (Eds.): Reasoning Web 2018, LNCS 11078, pp. 142–172, 2018.
https://doi.org/10.1007/978-3-030-00338-8_6

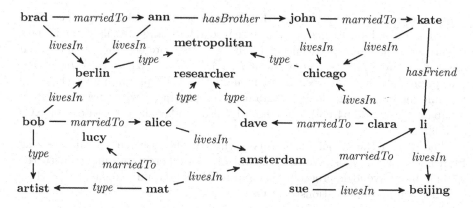

Fig. 1. Example KG: marriage relations and living places [76].

rather than false), the rules can be used to derive additional facts such as missing living places. Second, rules can be used to eliminate erroneous facts in KGs. For example, assuming that *livesIn* is a functional relation, a living place of a person could be questioned if it differs from the spouse's.

When rules are automatically learned, statistical measures like support, confidence and their variations are used to assess the quality of the rules. Most notably, the confidence of a rule is the fraction of facts predicted by the rule that are indeed in the KG. However, this is a meaningful measure for rule quality only when the KG is reasonably complete. For rules learned from incomplete KGs, confidence and other measures may be misleading, as they do not reflect the patterns in the missing facts. This might lead to the extraction of erroneous rules from incomplete and biased KGs. For example, a KG that stores a lot of information about families of popular scientists but lacks data in other domains, would yield a heavily biased rule r'_1:$hasChild(X, Y) \leftarrow worksAt(X, Z), educated(Y, Z)$, stating that workers of certain institutions often have children among the people educated there, as this is frequently the case for scientific families.

To address this issue, several rule measures that are specifically tailored towards incomplete KGs have been proposed [30,82] (see [3,18] for an overview of other measures). Along with KGs themselves, additional background knowledge could be used for better rule assessment. As proposed in [75] this includes metadata about the concrete numbers of facts of certain types that hold in the real world (e.g., *"Einstein has 3 children"*) automatically extracted from Web resources using techniques like [45]. Other types of background knowledge comprise description logic ontologies, e.g., [11] or more general hybrid theories, e.g., [36,43].

Horn rules such as r_1, might not always be sufficiently expressive to capture KG patterns accurately. Indeed, these rules cannot handle exceptions, which often appear in practice. For instance, application of r_1 mined from the KG in Fig. 1 results in the facts *livesIn(alice, berlin)*, *livesIn(dave, chicago)* and *livesIn(lucy, amsterdam)*. Observe that the first two facts might be suspected to be wrong. Indeed, both *alice* and *dave* are researchers, and the rule r_1 could have researcher as a potential exception resulting in its more accurate version given as $r_2 : livesIn(Y, Z) \leftarrow marriedTo(X, Y), livesIn(X, Z), not\ researcher(Y)$. Exception handling has been faced in inductive logic programming (ILP) by learning nonmonotonic logic programs, i.e., programs with negations from databases (e.g., [35,64,66]), and recently also studied in the context of KGs [29,76].

The aim of this article is to survey the current research on rule learning from knowledge graphs. We present and discuss different techniques with the roots in inductive logic programming and relational data mining as well as their interrelation and applications for KGs.

Tutorial Overview. In Sect. 2, we briefly introduce knowledge graphs and their key properties. We then provide necessary preliminaries on rule-based deductive reasoning over KGs and discuss the tasks within the area of inductive logic programming in Sect. 3. Section 4 describes recent research progress in the context of Horn rule induction for KG completion. We present techniques for nonmonotonic rule extraction in Sect. 5. Finally, in Sect. 6 we conclude the article with an outlook discussion, where we identify a number of promising directions for future work.

2 Knowledge Graphs

Knowledge graphs have been introduced in the Semantic Web community to create the "Web of data" that is readable by machines. They represent inter-linked collections of factual information, and are often encoded using the RDF data model [38]. This data model represents the content of a graph with a set of triples of the form ⟨*subject predicate object*⟩ corresponding to positive unary and binary first-order logic (FOL) facts.

Formally, we assume countable sets $\mathcal{R}$ of unary and binary predicate names (aka relations) and $\mathcal{C}$ of constants (aka entities). A knowledge graph $\mathcal{G}$ is a finite set of ground atoms of the form $p(s, o)$ and $c(s)$ over $\mathcal{R} \cup \mathcal{C}$. With $\Sigma_{\mathcal{G}} = \langle \mathcal{R}, \mathcal{C} \rangle$, the signature of $\mathcal{G}$, we denote elements of $\mathcal{R} \cup \mathcal{C}$ that occur in $\mathcal{G}$.

Example 1. Figure 1 shows a snippet of a graph about people, family and friend-ship relations among them as well as their living places and professions. For instance, the upper left part encodes the information that "Ann has a brother John, and lives with her husband Brad in Berlin, which is a metropolitan city" represented by the FOL facts {*hasBrother(ann, john)*, *livesIn(ann, berlin)*, *livesIn(brad, berlin)*, *metropolitan(berlin)*, *marriedTo(brad, ann)*}. The set $\mathcal{R}$ of relations appearing in the given KG contains the predicates *livesIn*, *marriedTo*, *hasBrother*, *hasFriend*, *researcher*, *metropolitan*, *artist*, while the set $\mathcal{C}$ of constants comprises of the names of people and locations depicted on Fig. 1.   □

Table 1. Examples of real-world KGs and their statistics [53,55].

Knowledge graphs	# Entities	# Relations	# Facts
DBpedia (en)	4.8 M	2800	176 M
Freebase	40 M	35000	637 M
YAGO3	3.6 M	76	61 M
Wikidata	46 M	4700	411 M
Google knowledge graph	570 M	35000	18000 M

All approaches for KG construction can be roughly classified into two major groups: manual and (semi-)automatic. The examples of KGs constructed manually include, e.g., WordNet [44] which has been created by a group of experts or Freebase [1] and Wikidata [77] which are constructed collaboratively by volunteers. Automatic population of KGs from semi-structured resources such as Wikipedia info-boxes using regular expressions and other techniques gave rise to, e.g., YAGO [72] and DBpedia [42]. There are also projects devoted to the extraction of facts from unstructured resources using natural language processing and machine learning methods. For example, NELL [47] and KnowledgeVault [16] belong to this category. Table 1 shows examples of some KGs and their statistics.

2.1 Incompleteness, Bias and Noise of Knowledge Graphs

While the existing KGs contain millions of facts, they are still far from being complete; Therefore they are treated under the open world assumption (OWA), i.e., facts not present in KGs are assumed to be unknown rather than false. For example, given only Germany as the living place of Albert Einstein, we cannot say whether $livedIn(einstein, us)$ is true or false. This is opposed to the closed world assumption (CWA) made in databases, under which $\neg livedIn(einstein, us)$ would be inferred.

Apart from incompleteness, both manually and semi-automatically created KGs also suffer from the problem of bias in the data. Indeed, manually created KGs such as Wikidata contain crowd-sourced information. While leveraging crowd-sourcing for KG construction, human curators from different countries add factual statements that they find interesting. Due to the difference in the population of contributors obviously facts about some countries are covered better than about others. For example, KGs typically store more facts about Austrians than Ghanians even though there are three times more inhabitants in Ghana than in Austria. Moreover, different contributors find different facts interesting, e.g., Austrians would add detailed information about music composers, while Ghanians about national athletes. This naturally leads to cultural bias in KGs. Likewise KGs that are semi-automatically extracted from Wikipedia infoboxes such as DBPedia and YAGO highly depend on the pre-defined properties that the infoboxes contain [40].

Along with completeness and absence of data bias, there are also other important aspects reflecting KGs' quality including their correctness. Regardless of the KG construction method, the resulting facts are rarely error-free. Indeed, in the case of manually constructed KGs human contributors might bring their own opinion on the added factual statements (e.g., Catalonia being a part of Spain or an independent country). On the other hand, automatically constructed KGs often contain noisy facts, since information extraction methods are imperfect. We refer the reader to [53,55] for further discussions on the available KGs and their quality.

The problems of KG completion and cleaning are among the central ones. Approaches for addressing them can be roughly divided into two groups: statistics-based, and logic-based. The firsts apply such techniques as tensor factorization, or neural-embedding-based models (see [54] for overview). The second group concentrates on logical rule learning [28]. In this tutorial, we primarily focus on rule-based techniques, and their application for the KG completion task. In the following, we assume that the given KG $\mathcal{G}$ stores only a subset of all true facts.

Suppose we had an *ideal* KG $\mathcal{G}^{i1}$ that contains *all* correct and true facts in the world reflecting the relations from $\mathcal{R}$ that hold among the entities in $\mathcal{C}$. The gap between $\mathcal{G}$ and its ideal version $\mathcal{G}^i$ is defined as follows:

Definition 1 (Incomplete Knowledge Graph [12]). *An incomplete knowledge graph is a pair $G = (\mathcal{G}, \mathcal{G}^i)$ of two KGs, where $\mathcal{G} \subseteq \mathcal{G}^i$ and $\Sigma_{\mathcal{G}} = \Sigma_{\mathcal{G}^i}$. We call $\mathcal{G}$ the available graph and $\mathcal{G}^i$ the ideal graph.*

Note that $\mathcal{G}^i$ is an abstract construct, which is normally unavailable. Intuitively, *rule-based KG completion task* concerns the reconstruction of the ideal KG (or its approximation) by means of rules induced from the available KG and possibly other external information sources.

3   Rules and Reasoning

In this section, we briefly review the concepts of rules, logic programming (see [21] for more details) as well as deductive and inductive reasoning.

3.1   Logic Programs

Logic programs consist of a set of rules. Intuitively, a rule is an if-then expression, whose if-part may contain several conditions, some possibly with negation. The then-part has a single atom that has to hold, whenever the if-part holds. In general, the then-part can also contain disjunctions, but in this tutorial we consider only non-disjunctive rules. More formally,

---

[1] The superscript i stands for *ideal*.

Definition 2 (Rule). *A rule* r *is an expression of the form*

$$h(\boldsymbol{X}) \leftarrow b_1(\boldsymbol{Y_1}), \ldots, b_k(\boldsymbol{Y_k}), not\ b_{k+1}(\boldsymbol{Y_{k+1}}), \ldots, not\ b_n(\boldsymbol{Y_n}) \qquad (1)$$

where $h(\boldsymbol{X}), b_1(\boldsymbol{Y_1}), \ldots, b_n(\boldsymbol{Y_n})$ *are first-order atoms and the right-hand side of the rule is a conjunction of atoms. Moreover,* $\boldsymbol{X}, \boldsymbol{Y_1}, \ldots, \boldsymbol{Y_n}$ *are tuples of either variables or constants whose length corresponds to the arity of the predicates* $h, b_1, \ldots, b_n$ *respectively.*

The left-hand side of a rule r is referred to as its head, denoted by $head(r)$, while the right-hand side is its body $body(r)$. The positive and negative parts of the body are respectively denoted as $body^+(r)$ and $body^-(r)$. A rule r is called *positive* or *Horn* if $body^-(r) = \emptyset$.

For simplicity, in this work we use the shortcut

$$H \leftarrow B, not\ E \qquad (2)$$

to denote rules with a single negated atom, i.e., here $H = h(\boldsymbol{X}), B = b_1(\boldsymbol{Y_1}), \ldots, b_k(\boldsymbol{Y_k}), E = b_{k+1}(\boldsymbol{Y_{k+1}})$.

Example 2. Consider r_2 from Sect. 1. We have that $head(r_2) = \{livesIn(Y, Z)\}$, while $body^+(r_2) = \{isMarriedTo(X, Y), livesIn(X, Z)\}$, and moreover it holds that, $body^-(r_2) = \{not\ researcher(Y)\}$. $\square$

A logic program P is *ground* if it consists of only ground rules, i.e. rules without variables.

Example 3. For instance, a possible grounding of the rule r_2 is given as follows
$$livesIn(dave, chicago) \leftarrow livesIn(clara, chicago), isMarriedTo(clara, dave),$$
$$not\ researcher(dave). \qquad \square$$

Ground instantiation $Gr(P)$ of a nonground program P is obtained by substituting variables with constants in all possible ways.

Definition 3 (Herbrand Universe, Base, Interpretation). *A* Herbrand universe $HU(P)$ *is a set of all constants occurring in the given program* P. *A* Herbrand base $HB(P)$ *is a set of all possible ground atoms that can be formed with predicates and constants appearing in* P. *A* Herbrand interpretation *is any subset of* $HB(P)$.

We now formally define the satisfaction relation.

Definition 4 (Satisfaction, Model). *An interpretation* I *satisfies*

- *a ground atom* a, *denoted* $I \models a$, *if* $a \in I$,
- *a negated ground atom* $not\ a$, *denoted* $I \models not\ a$, *if* $I \not\models a$,
- *a conjunction* $b_1, \ldots, b_n$ *of ground literals, denoted* $I \models b_1, \ldots, b_n$, *if for each* $i \in \{1, \ldots, n\}$ *it holds that* $I \models b_i$,
- *a ground rule* r, *denoted* $I \models r$ *if* $I \models body(r)$ *implies* $I \models head(r)$, *i.e., if all literals in the body hold then the literal in the head also holds.*

An interpretation I is a *model* of a ground program P, if $I \models r$ for each rule $r \in P$. A model I is *minimal* if there is no other model $I' \subset I$. By $MM(P)$ we denote the set-inclusion minimal model of a ground positive program P.

The classical definition of answer sets based on the Gelfond-Lifschitz reduct [31] is given as follows.

Definition 5 (Gelfond-Lifschitz Reduct, Answer Set [31]). *An interpretation I of P is an* answer set *(or* stable model*) of P iff $I \in MM(P^I)$, where P^I is the* Gelfond-Lifschitz (GL) reduct *of P, obtained from $Gr(P)$ by removing (i) each rule r such that $Body^-(r) \cap I \neq \emptyset$, and (ii) all the negative atoms from the remaining rules. The set of answer sets of a program P is denoted by $AS(P)$.*

An alternative definition of answer sets relies on the FLP-reduct [25] which might be more intuitive. For the class of programs considered in this work, the GL-reduct and FLP-reduct are equivalent.

Definition 6 (Faber-Leone-Pfeifer Reduct, Answer Set [25]). *An interpretation I of P is an* answer set *(or* stable model*) of P iff $I \in MM(fP^I)$, where fP^I is the* Faber-Leone-Pfeifer (FLP) reduct *of P, obtained from $Gr(P)$ by keeping only rules r, whose bodies are satisfied by I, i.e.,*

$$fP^I = \{r \in P \mid head(r) \leftarrow body(r), I \models body(r)\}.$$

Example 4. Consider the program

$$P = \left\{ \begin{array}{l} (1) \ \textit{livesIn}(\textit{brad}, \textit{berlin}); \ (2) \ \textit{isMarriedTo}(\textit{brad}, \textit{ann}); \\ (3) \ \textit{livesIn}(Y, Z) \leftarrow \textit{isMarriedTo}(X, Y), \textit{livesIn}(X, Z), \textit{not} \ \textit{researcher}(Y) \end{array} \right\}$$

The relevant part of ground instantiation $Gr(P)$ of P is obtained by substituting X, Y, Z with *brad, ann* and *berlin* respectively. For $I = \{\textit{isMarriedTo}(\textit{brad},\textit{ann}),\textit{livesIn}(\textit{ann},\textit{berlin}), \textit{livesIn}(\textit{brad}, \textit{berlin})\}$, both the GL- and FLP-reduct contain $\textit{livesIn}(\textit{ann}, \textit{berlin}) \leftarrow \textit{livesIn}(\textit{brad}, \textit{berlin})$, $\textit{isMarriedTo}(\textit{brad}, \textit{ann})$ and the facts (1), (2). As I is a minimal model of these reducts, I is an answer set of P.   □

Apart from the model computation, another important task concerns deciding whether a given atom a is entailed from the program P denoted by $P \models a$. Typically, one distinguishes between brave and cautious entailment. We say that an atom a is *cautiously entailed* from a program P ($P \models_c a$) if a is present in all answer sets of P. Conversely, an atom a is *bravely entailed* from a program P ($P \models_b a$) if it is present in at least one answer set of P. If a program has only a single answer set, e.g., it is positive, then obviously cautious and brave entailments coincide, in which case we omit the subscript.

Example 5. For P from Example 4 and $a = \textit{livesIn}(\textit{ann}, \textit{berlin})$, we have that a is entailed both bravely and cautiously from P, i.e., $P \models a$, while for $a' = \textit{marriedTo}(\textit{brad}, \textit{brad})$, it holds that $P \not\models a'$.   □

The answer set semantics for nonmonotonic logic programs is based on the CWA, under which whatever can not be derived from a program is assumed to be false. Nonmonotonic logic programs are widely applied for formalizing common sense deductive reasoning over incomplete information.

3.2 Inductive Reasoning Tasks

So far we have focused on logic programs and their semantics for deductive reasoning. We now discuss the problem of automatic rule induction, which is a research area commonly referred to as rule learning (see, e.g., [27,58]).

Broadly speaking, rule learning is an important sub-field of machine learning, which focuses on symbolic methods for intelligent data analysis, i.e., methods that employ a certain description language in which the learned knowledge is represented.

First-order learning approaches are also referred to as *inductive logic programming (ILP)*, since the patterns they discover are expressed in relational formalisms of first-order logic (see [58] for overview). The goal of ILP is to generalize individual instances/observations in the presence of background knowledge by building hypotheses about yet unseen instances. The most commonly addressed task in ILP is the task of learning logical definitions of relations. From training examples ILP induces a logic program (predicate definition) corresponding to a view that defines the target relation in terms of other relations that are given as background knowledge.

More formally, the classical inductive logic programming task of learning from positive and negative examples (also known as learning from entailment) is defined as follows:

Definition 7 (Inductive Learning from Examples [49])

Given:
- *Positive examples E^+ and negative examples E^- over the target n-ary relation p, i.e. sets of facts;*
- *Background knowledge T, i.e. a set of facts over various relations and possibly rules that can be used to induce the definition of p;*
- *Syntactic restrictions on the definition of p.*

Find:
- *A hypothesis Hyp that defines the target relation p, which is (i) complete, i.e., $\forall e \in E^+$, it holds that $T \cup Hyp \models e$, and (ii) consistent, i.e., $\forall e' \in E^- : T \cup Hyp \not\models e'$.*

Example 6. Suppose that you possess information about some of the relationships between people in your family and their genders. However, you do not know what the relationship *fatherOf* actually means. You might have the following beliefs, i.e., background knowledge.

$$T = \left\{ \begin{array}{l} (1)\ parentOf(john, mary); (2)\ male(john); (3)\ parentOf(david, steve); \\ (4)\ male(david); (5)\ parentOf(kathy, ellen); (6)\ female(kathy); \end{array} \right\}$$

Moreover, you are given the following positive and negative examples.

Table 2. Overview of classical ILP systems.

System	Input data	Output rules	Multiple predicates	Increment	Interact	Noise handling
FOIL [57]	Examples	Horn	No	No	No	Yes
GOLEM [51]	Examples	Horn	No	No	No	Yes
LINUS [19]	Examples	Horn	No	No	No	Yes
ALEPH [71]	Examples	Horn	No	No	Yes	Yes
MPL [61]	Examples	Horn	Yes	No	No	No
MIS [69]	Examples	Horn	Yes	Yes	Yes	No
MOBAL [48]	Examples	Horn	Yes	Yes	No	No
CIGOL [50]	Examples	Horn	Yes	Yes	Yes	No
CLINT [13]	Examples	Horn	Yes	Yes	Yes	No
σILP [24]	Examples	Horn	No	No	No	Yes
DROPS [8]	Examples	NM	No	No	No	No
ASPAL [9]	Examples	NM	No	No	No	No
XHAIL [64]	Examples	NM	No	No	No	No
ILASP [41]	Interpretations	NM	Yes	No	No	No
ILED [37]	Examples	NM	No	Yes	No	No

$$E^+ = \{fatherOf(john, mary), fatherOf(david, steve)\}$$
$$E^- = \{fatherOf(kathy, ellen), fatherOf(john, steve)\}$$

One of the possible hypothesis that can be induced from the above knowledge reflecting the meaning of the *fatherOf* relation is given as follows:

$Hyp : fatherOf(X, Y) \leftarrow parentOf(X, Y), male(X)$. This hypothesis is consistent with the background theory T, and together with T it entails all of the positive examples, and none of the negative ones. The classical ILP task concerns automatic extraction of such hypothesis. □

In an alternative setting, known as *learning from interpretations*, instead of positive and negative examples over a certain relation, one is given a Herbrand interpretation I, i.e., a set of facts, and conditions (i) and (ii) in Definition 7 are replaced with the requirement that I is a minimal model of $Hyp \cup T$. Formally,

Definition 8 (Inductive Learning from Interpretations [60])

Given:
- *An interpretation I, i.e., a set of facts over various relations*
- *Background knowledge T, i.e., a set facts and possibly rules*
- *Syntactic restrictions on the form of rules to be induced*

Find:
- *A hypothesis Hyp, such that I is a minimal Herbrand model of $Hyp \cup T$.*

Several variations of learning from interpretations task have been studied including learning from answer sets [41, 66], where given possibly multiple partial interpretations, the goal is to find a logic program, which has extensions of (all of) the provided interpretations as answer sets.

To date, the main tasks considered in the ILP area can be classified based on the following parameters [4, 67].

- *type of the data source*, e.g., positive/negative examples, interpretations, text
- *type of the output knowledge*, e.g., Horn/nonmonotonic rules over single or multiple predicates, description logic (DL) class descriptions, class inclusions
- *the way the data is given as input*, e.g., all data at once or incrementally
- *availability of an oracle*, e.g., involvement of a human expert in the loop
- *quality of the data source*, e.g., noisy or clean
- *data (in)completeness assumption*, e.g., OWA, CWA
- *availability and type of background knowledge*, e.g., DL ontology, set of datalog rules, hybrid theories, etc.

An overview of some of the systems for Horn and nonmonotonic (NM) rule induction with their selected properties is presented in Table 2.

4   Rule Learning for Knowledge Graph Completion

The majority of the classical existing rule induction methods mentioned in Sect. 3 assume that the given data from which the rules are induced is complete, accurate and representative. Therefore, they rely on CWA and aim at extracting rule hypotheses that perfectly satisfy the criteria from Definition 7 or Definition 8. On the other hand, as discussed in Sect. 2.1, knowledge graphs are highly incomplete, noisy and biased. Moreover, the real world is very complicated, and its exact representation often cannot be acquired from the data, meaning that the task of inducing a perfect rule set from a KG is typically unfeasible. Therefore, in the context of KGs, one normally aims at extracting certain regularities from the data, which are not universally correct, but when seen as rules predict a sufficient portion of true facts.

In other words, the goal of automatic rule-based KG completion is to learn a set R of logic rules from the available graph $\mathcal{G}$, such that their application on $\mathcal{G}$ results in a good approximation of the ideal graph $\mathcal{G}^i$. More formally,

Definition 9 (Rule-based KG Completion). *Let $\mathcal{G}$ be a KG over the signature $\Sigma_{\mathcal{G}} = \langle \mathcal{R}, \mathcal{C} \rangle$. Let, moreover, R be a set of rules with predicates from $\mathcal{R}$ induced from $\mathcal{G}$. Then rule-based completion of $\mathcal{G}$ w.r.t R is a graph $\mathcal{G}_R$ constructed from any answer set $\mathcal{G}_R \in AS(R \cup \mathcal{G})$.*

Example 7. Given the KG in Fig. 1 as $\mathcal{G}$ and the rule set $R = \{r_2\}$, where r_2 is from Sect. 1 we have $\mathcal{G}_R = \mathcal{G} \cup \{livesIn(lucy, amsterdam)\}$. $\qquad\qquad\square$

If the ideal graph $\mathcal{G}^i$ was known, then the problem of rule-based KG completion would be essentially the same as the task of learning from interpretations given in Definition 8, with $\mathcal{G}$ playing a role of the background knowledge T and $\mathcal{G}^i$ the interpretation I. However, unfortunately $\mathcal{G}^i$ is unknown, and it cannot be easily reconstructed, since KGs follow OWA.

Moreover, reusing the methods that induce logical theories from a set of positive and negative examples from Definition 7 is likewise not possible due to the following important obstacles [76].

First, the target predicates (e.g. *fatherOf* from Example 6) can not be easily identified, since we do not know which parts of the considered KG need to be completed. A standard way of addressing this issue would be just to learn rules for all the different predicate names occurring in the KG. Unfortunately, this is unfeasible given the huge size of KGs. Second, the negative examples are not available, and they cannot be easily obtained from, e.g., domain experts due to - once again - the huge size of KGs.

To overcome the above obstacles, it is more appropriate to treat the KG completion problem as an unsupervised relational learning task [30]. In this section, we describe approaches that rely on *relational association rule learning* techniques for extraction of *Horn rules* from incomplete KGs. These concern the discovery of frequent patterns from a data set and their subsequent transformation into rules (see, e.g., [15] as the seminal work in this direction).

First, we describe the relational association rules, and how they are traditionally evaluated under CWA. Then, we present the standard relational rule learning techniques, which usually proceed in two steps: rule construction and rule assessment.

4.1   Relational Association Rules

An association rule is a rule where certain properties of the data in the body of the rule are related to other properties in its head. For an example of an association rule, consider a database containing transactional data from a store selling computer equipment. From this data one can extract the association rule stating that 70% of the customers buying a laptop also buy a docking station. The knowledge that such rule reflects can assist in planning store layout or deciding which customers are likely to respond to an offer.

Traditionally, the discovery of association rules has been performed on data stored in a single table. Recently, however, many methods for mining relational, i.e., graph-based data have been proposed [58].

The notion of multi-relational association rules is heavily based on frequent conjunctive queries and query subsumption [32].

Definition 10 (Conjunctive Query). *A conjunctive query Q over $\mathcal{G}$ is of the form $p_1(\mathbf{X_1}), \ldots, p_m(\mathbf{X_m})$, where $\mathbf{X_i}$ are symbolic variables or constants and $p_i \in \mathcal{R}$ are unary or binary predicates. The* answer *of Q on $\mathcal{G}$ is the set $Q(\mathcal{G}) = \{(\nu(\mathbf{X_1}), \ldots, \nu(\mathbf{X_m})) \mid \forall i : p_i(\nu(X_i), \nu(Y_i)) \in \mathcal{G}\}$ where ν is a function that maps variables and constants to elements of $\mathcal{C}$.*

The *(absolute) support* of a conjunctive query Q in a KG $\mathcal{G}$, is the number of distinct tuples in the answer of Q on $\mathcal{G}$ [15].

Example 8. The support of the query

$$Q(X, Y, Z) :\text{-} \; marriedTo(X, Y), livesIn(X, Z)$$

over $\mathcal{G}$ in Fig. 1 asking for people, their spouses and living places is 6. □

Definition 11 (Association Rule). *An* association rule *is of the form* $Q_1 \Rightarrow Q_2$, *such that* Q_1 *and* Q_2 *are both conjunctive queries, and the body of* Q_1 *considered as a set of atoms is included in the body of* Q_2, *i.e.,* $Q_1(\mathcal{G}') \subseteq Q_2(\mathcal{G}')$ *for any possible KG* $\mathcal{G}'$.

Example 9. For instance, from the above $Q(X, Y, Z)$ and

$$Q'(X, Y, Z) :\text{-} \; marriedTo(X, Y), \; livesIn(X, Z), livesIn(Y, Z)$$

we can construct the association rule $Q \Rightarrow Q'$. □

Association rules are sometimes exploited for reasoning purposes, and thus (with some abuse of notation) can be treated as logical rules, i.e., for $Q_1 \Rightarrow Q_2$ we write $Q_2 \backslash Q_1 \leftarrow Q_1$, where $Q_2 \backslash Q_1$ refers to the set difference between Q_2 and Q_1 considered as sets of atoms. For example, $Q \Rightarrow Q'$ from above corresponds to *r1* from Sect. 1.

A large number of measures for evaluating the quality of association rules and their subsequent ranking have been proposed, e.g., *support, confidence*.

For $r : \quad H \leftarrow B, not \; E$ (see Eq. 2), with $H = h(X, Y)$, B, E involving variables from $\mathbf{Z} \supseteq \{X, Y\}$ and a KG $\mathcal{G}$, the *(standard) confidence* is given as:

$$conf(r, \mathcal{G}) = \frac{r\text{-}supp(r, \mathcal{G})}{b\text{-}supp(r, \mathcal{G})}$$

where $r\text{-}supp(r, \mathcal{G})$ and $b\text{-}supp(r, \mathcal{G})$ are the *rule support* and *body support*, respectively, which are defined as follows:

$$r\text{-}supp(r, \mathcal{G}) = \#(x, y) : h(x, y) \in \mathcal{G}, \exists \mathbf{z} \; B \in \mathcal{G}, E \notin \mathcal{G}$$
$$b\text{-}supp(r, \mathcal{G}) = \#(x, y) : \exists \mathbf{z} \; B \in \mathcal{G}, E \notin \mathcal{G}$$

Example 10. Consider the rules r_1, r_2, and the KG $\mathcal{G}$ in Fig. 1, we have $r\text{-}supp(r_1, \mathcal{G}) = r\text{-}supp(r_2, \mathcal{G}) = 3$, $b\text{-}supp(r_1, \mathcal{G}) = 6$ and $b\text{-}supp(r_2, \mathcal{G}) = 4$. Hence, $conf(r_1, \mathcal{G}) = \frac{3}{6}$ and $conf(r_2, \mathcal{G}) = \frac{3}{4}$. □

Another popular metrics, which is shown to guarantee the high predictive power [3] by measuring the intensity of rule's implication [18] is *conviction*, defined as:

$$conv(r, \mathcal{G}) = \frac{1 - rel\text{-}supp(H, \mathcal{G})}{1 - conf(r, \mathcal{G})}$$

where $rel\text{-}supp(H, \mathcal{G})$ is the relative support of the head, measured by:

$$rel\text{-}supp(H, \mathcal{G}) = \frac{\#(x, y) : h(x, y) \in \mathcal{G}}{(\#x : \exists y : h(x, y) \in \mathcal{G}) \times (\#y : \exists x : h(x, y) \in \mathcal{G})}$$

Example 11. For the KG in Fig. 1, we have $rel\text{-}supp(livesIn, \mathcal{G}) = \frac{10}{10 \times 4} = \frac{1}{4}$, thus $conv(r_1, \mathcal{G}) = \frac{1-\frac{1}{4}}{1-\frac{3}{6}} = \frac{3}{2}$ and $conv(r_2, \mathcal{G}) = \frac{1-\frac{1}{4}}{1-\frac{3}{4}} = 3$. □

4.2   Rule Construction

We now briefly summarize some of the state-of-the-art methods for rule construction, most of which extract so-called *closed* rules, i.e., rules, in which every variable appears at least twice. Restriction to closed rules ensures the actual prediction of a fact by a rule, but not just its existence [30].

Example 12. The rules r_1 and r_2 from Sect. 1 are *closed*. An example of a *non-closed* rule is:

$$\exists Z \; livesIn(Y, Z) \leftarrow isMarriedTo(X, Y)$$

which states that married people live somewhere. This rule cannot infer the exact living place of a person, but merely its existence, and thus it is less interesting in this context. □

The most prominent examples of systems that are specifically tailored towards inducing Horn rules from KGs are AMIE [30] and RDF2Rules [79].

AMIE. AMIE [30] is a state-of-the-art Horn rule mining system. Apart from a KG, it expects the maximum length of the rule and the threshold value for its support. In AMIE, rules are treated as sequences of atoms, where the first atom is the head, and subsequent atoms form the body of the rule. The algorithm maintains a queue of intermediate rules, which initially stores a single rule with an empty body for every KG relation. Rules are removed from the queue and refined by adding literals to the body according to a language bias that specifies allowed rule forms (e.g., based on the user-provided rule length). The system then estimates the support of the rule, and if it exceeds the given threshold, the rule is output to the user and also added to the queue for possible further processing. Refinement of a rule relies on the following set of mining operators used to extend the sequences of atoms in the rule body:

- *add dangling atom*: add a new positive atom with one fresh variable, i.e., variable not appearing elsewhere in the rule;
- *add instantiated atom*: add a positive atom with one argument being a constant and the other one being a shared variable, i.e., variable already present in another rule atom;
- *add closing atom*: add a positive atom with both of its arguments being shared variables.

The implementation of AMIE employs a variety of techniques from the database area, which allow it to achieve high scalability.

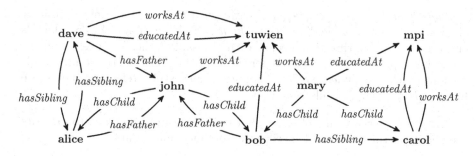

Fig. 2. Example KG: family relations, working and education places [75].

RDF2Rules. While AMIE mines a single rule at a time, RDF2Rules [79] parallelizes this process by extracting *frequent predicate cycles* (FPCs) of a certain length k, which have the following form:

$$(X_1, p_1^{d_1}, X_2, p_2^{d_2}, \ldots, X_k, p_k^{d_k}, X_1)$$

where, X_is are variables to appear in the extracted rules, p_is are predicates connecting these variables, and d_is $\in \{0, 1\}$ reflect the direction of the edges in the KG labeled by the respective predicates. To extract FPCs, the RDF2Rules algorithm first mines the *frequent predicate paths* (FPPs) of the form:

$$(X_1, p_1^{d_1}, X_2, p_2^{d_2}, \ldots, X_k, p_k^{d_k}, X_{k+1})$$

of the length k, which are obtained recursively based on FPPs of the length $k-1$. FPCs are then created from FPPs by merging the last variable X_{k+1} with the first one X_1. After FPCs are mined, rules are extracted from them by choosing a single predicate to be in the rule head, and collecting the rest into it's body.

RDF2Rules is capable of accounting for unary predicates (i.e., types of entities), which are neglected in AMIE for scalability reasons. The unary predicates are added to the constructed rule at the final stage after analyzing the frequent types for FPCs corresponding to a given rule. While RDF2Rules performs the rule extraction faster than AMIE due to an effective pruning strategy used in the process of mining FPCs, the supported rule patterns are more restrictive.

4.3   Rule Evaluation

Most of state-of-the-art KG-based positive rule mining systems differ from each other with respect to the employed rule ranking function. The ranking metrics from data mining such as support and confidence (see e.g., [18] for overview of others) presented in Sect. 4.1 have been designed for datasets treated under the CWA, and they can be counterintuitive for the KGs, in which facts are largely missing.

Example 13. For instance, consider the KG $\mathcal{G}'$ in Fig. 2 [75], which presents information about scientific families. The heavily biased rule from Sect. 1:

$r'_1 : hasChild(X, Y) \leftarrow worksAt(X, Z), educated(Y, Z)$ can be mined from it along with $r'_2 : hasSibling(X, Z) \leftarrow hasFather(X, Y), hasChild(Y, Z)$, which is an accurate rule stating that people with the same father are likely siblings. Since the graph is highly incomplete for the *hasSibling* relation, the standard rule measures such as confidence reflect a counterintuitive rule preference. Indeed, we have $conf(r'_1, \mathcal{G}') = \frac{2}{8}$, while $conf(r'_2, \mathcal{G}') = \frac{1}{6}$. □

Below we present some of other alternative measures, designed to quantify the quality of rules extracted specifically from incomplete data.

PCA Confidence. In [30], a completeness-aware rule scoring based on the partial completeness assumption (PCA) has been introduced. The idea of PCA is that whenever at least one object for a given subject and a predicate is in the KG (e.g., "Eduard is Einstein's child"), then all objects for that subject-predicate pair (Einstein's children) are assumed to be known. PCA relies on a hypothesis that the data is usually added to KGs in batches, i.e., if at least one child for a person has been added, then most probably all person's children are present in the KG. This assumption has turned out to be indeed valid in real-world KGs for some topics [30]. The PCA confidence is defined as follows:

$$conf_{pca}(r, \mathcal{G}) = \frac{r\text{-}supp(r, \mathcal{G})}{\#(x, y) : \exists z : B \in \mathcal{G} \land \exists y' : h(x, y') \in \mathcal{G}}$$

However, the effectiveness of the PCA confidence decreases when applied on highly incomplete KGs as experienced in [34].

Example 14. Given the rules r'_1 and r'_2 from Example 13 and the KG from Fig. 2, we have that $conf_{pca}(r'_1, \mathcal{G}') = \frac{4}{2}$. Indeed, since *carol* and *dave* are not known to have any children, four existing body substitutions are not counted in the denominator. Meanwhile, we have $conf_{pca}(r'_2, \mathcal{G}') = \frac{1}{6}$, since all people that are predicted to have siblings by r'_2 already have siblings in the available graph.

For the KG in Fig. 1 and the earlier introduced rule r_1, it holds that $conf_{pca}(r_1, \mathcal{G}) = \frac{3}{4}$, since we do not know any living places of *lucy* and *dave*. □

Soft Confidence. *Soft confidence* measure introduced in [79], is also designed to work under OWA, and for a rule $r : h(X, Y) \leftarrow B$ it is formally defined as follows:

$$conf_{st}(r, \mathcal{G}) = \frac{r\text{-}supp(r, \mathcal{G})}{b\text{-}supp(r, \mathcal{G}) - \sum_{x \in U^r} Pr(x, h, \mathcal{G})}$$

where U^r is the set of entities that have no outgoing h-edges in $\mathcal{G}$, but do have some in $\mathcal{G}_r$, and $Pr(x, h, \mathcal{G})$ is the probability of the entity x having the relation h, approximated using entity *type* information [79]. More specifically,

$$Pr(x, h, \mathcal{G}) = max_{t \in T_x} \frac{|Inst_h(t, \mathcal{G})|}{|Inst(t, \mathcal{G})|}$$

where $T_x = \{t \mid t(x) \in \mathcal{G}\}$, $Inst(t, \mathcal{G}) = \{x' \mid t(x') \in \mathcal{G}\}$, and, moreover, $Inst_h(t, \mathcal{G}) = \{x' \in Inst(t, \mathcal{G}) \mid \exists x'' : h(x', x'') \in \mathcal{G}\}$. Intuitively, soft confidence is designed to avoid the under-fitting of standard confidence and over-fitting of PCA confidence by accounting for the probability of entities having certain relations.

Example 15. Consider the KG $\mathcal{G}$ from Fig. 1, we have $U^{r_1} = \{lucy, dave\}$. Moreover, $Pr(dave, livesIn, \mathcal{G}) = \frac{1}{2}$, since *dave* has only 1 type *researcher*, and in total the KG stores 2 researchers (*dave, alice*) but only *alice*'s living place (*amsterdam*) is known to $\mathcal{G}$. In contrast, $Pr(lucy, livesIn, \mathcal{G}) = 0$, as *lucy* has no *type* information. Based on these numbers, we have $conf_{st}(r_1, \mathcal{G}) = \frac{3}{6 - \frac{1}{2}} = \frac{6}{11}$.

□

RC Confidence. The authors of [82] have recently proposed the *RC confidence* as an attempt to rely on assumptions about the tuples not in the KG when evaluating the quality of a potential rule. The intuition behind *RC confidence* is that for computing rule's confidence one does not necessarily have to know which among rule predictions are true; just estimating their number is sufficient. The key assumption for such estimation is that the proportion of positive facts covered by a rule is the same for both true and unknown facts, which is reflected in the following *relationship coverage* defined for $r : h(X, Y) \leftarrow B$ as follows

$$\frac{r\text{-}supp(r, \mathcal{G})}{\#(x, y) : h(x, y) \in \mathcal{G}} = \frac{|UP(r, \mathcal{G})|}{\#(x, y) : h(x, y) \in \mathcal{G}^i \backslash \mathcal{G}}$$

where $UP(r, \mathcal{G}) = \mathcal{G}_r \cap \mathcal{G}^i$ (assumed to be 0 in the case of standard confidence). Intuitively, this equation would hold if the instantiations of r that appear in $\mathcal{G}$ were selected completely at random [23] from the set of all true facts for r.

To approximately determine the size of $UP(r, \mathcal{G})$ the authors rely on the proportion β of all the unlabeled examples that are true in $\mathcal{G}^i$, i.e., $UP(r, \mathcal{G}) = \beta * \#(x, y) : h(x, y) \notin \mathcal{G}$, and propose several ways for calculating β both empirically via sampling and theoretically based on properties of $\mathcal{G}$.

Finally, provided that there is a way to estimate $|UP(r, \mathcal{G})|$, the following formula is used for computing the *RC-confidence*:

$$conf_{rc}(r, \mathcal{G}) = \frac{r\text{-}supp(r, \mathcal{G}) + |UP(r, G)|}{b\text{-}supp(r, \mathcal{G})}$$

(In)completeness Metadata. In the solutions for the rule-based KG completion problem discussed so far, no external meta-information from outside of the KG about potential existence of certain types of facts was exploited. However, this knowledge is obviously useful, and furthermore, it can even be extracted from the Web in the form of cardinality statements, e.g., Brad has three children. If a given KG mentions just a single Brad's child, we could aim at extracting rules that predict the missing one.

Based on this intuition, the recent work on completeness-aware rule learning (CARL) [75] proposed improvements of rule scoring functions by making use of this additional (in)completeness metadata.

In particular, such metadata is presented using cardinality statements by reporting (the numerical restriction on) the absolute number of facts over a certain relation in the ideal graph $\mathcal{G}^i$. More specifically, the partial function num is defined that takes as input a predicate p and an entity x and outputs a natural number corresponding to the number of facts in $\mathcal{G}^i$ over p with x as the first argument:

$$num(p, x) = \#y : p(x, y) \in \mathcal{G}^i \qquad (3)$$

These cardinality statements can be obtained via Web extraction techniques [46]. It is possible to rewrite cardinalities on the number of subjects for a given predicate and object with such statements provided that inverse relations can be expressed in a KG. Naturally, the number of missing facts for a given p and x can be obtained as

$$miss(p, x, \mathcal{G}) = num(p, x) - \#y : p(x, y) \in \mathcal{G}$$

In the CARL framework, given a KG and its related cardinality statements of the above form, two indicators are defined for a given rule $r : h(X, Y) \leftarrow B$, reflecting the number of new predictions made by r in incomplete (npi) and, respectively, complete (npc) KG parts:

$$npi(r, \mathcal{G}) = \sum_x min(\#y : h(x, y) \in \mathcal{G}_r \backslash \mathcal{G}, miss(h, x, \mathcal{G}))$$

$$npc(r, \mathcal{G}) = \sum_x max(\#y : h(x, y) \in \mathcal{G}_r \backslash \mathcal{G} - miss(h, x, \mathcal{G}), 0)$$

Using these indicators, a class of **completeness-aware rule measures** have been defined in [75], which we briefly present next.

Completeness Confidence. First, incompleteness information is used to determine whether to consider an instance in the unknown part of the rule as a counterexample or not. Formally, the *completeness confidence* is defined as:

$$conf_{comp}(r, \mathcal{G}) = \frac{r\text{-}supp(r, \mathcal{G})}{b\text{-}supp(r, \mathcal{G}) - npi(r, \mathcal{G})}$$

Example 16. Consider $\mathcal{G}'$ in Fig. 2 and cardinality statements for it:

$num(hC, john) = num(hC, mary) = 3; num(hC, alice) = 1;$
$num(hC, carol) = num(hC, dave) = 0;$
$num(hS, alice) = num(hS, carol) = num(hS, dave) = 2;$
$num(hS, bob) = 3;$

where hC, hS stand for *hasChild* and *hasSibling*, respectively. We have:

$miss(hC, mary, \mathcal{G}') = miss(hC, john, \mathcal{G}') = miss(hC, alice, \mathcal{G}') = 1;$

$miss(hC, carol, \mathcal{G}') = miss(hC, dave, \mathcal{G}') = 0$;
$miss(hS, bob, \mathcal{G}') = miss(hS, carol, \mathcal{G}') = 2$;
$miss(hS, alice, \mathcal{G}') = miss(hS, dave, \mathcal{G}') = 1$;
For the rules r'_1 and r'_2 from Example 13 we have $conf_{comp}(r'_1, \mathcal{G}') = \frac{2}{6}$ and $conf_{comp}(r'_2, \mathcal{G}') = \frac{1}{2}$, which establishes the desired rule ranking.   □

Completeness Precision and Recall. In the spirit of information retrieval, the notions of *completeness precision* and *completeness recall* are defined to measure the rule quality based on its predictions in complete and incomplete KG parts:

$$precision_{comp}(r, \mathcal{G}) = 1 - \frac{npc(r, \mathcal{G})}{b\text{-}supp(r, \mathcal{G})}, \quad recall_{comp}(r, \mathcal{G}) = \frac{npi(r, \mathcal{G})}{\sum_x miss(h, x, \mathcal{G})}$$

Intuitively, rules having high precision are rules that predict few facts in complete parts, while rules having high recall are rules that predict many facts in incomplete ones. Rule scoring could also be based on any weighted combination of these two metrics.

Example 17. We have $npi(r'_1, \mathcal{G}') = 2$, $npc(r'_1, \mathcal{G}') = 4$, while $npi(r'_2, \mathcal{G}') = 4$, $npc(r'_2, \mathcal{G}') = 1$, resulting in $precision_{comp}(r'_1, \mathcal{G}') = 0.5$, $recall_{comp}(r'_1, \mathcal{G}') \approx 0.67$, and $precision_{comp}(r'_2, \mathcal{G}') \approx 0.83$, $recall_{comp}(r'_2, \mathcal{G}') \approx 0.67$.   □

Directional Metric. If rule mining does make use of completeness information, and both do not exhibit any statistical bias, then intuitively the rule predictions and the (in)complete areas should be statistically independent. On the other hand, correlation between the two indicates that the rule-mining is *(in)completeness-aware.* Following this intuition, the *directional metric* has been proposed, which measures the proportion between predictions in complete and incomplete parts as follows:

$$dm(r, \mathcal{G}) = \frac{npi(r, \mathcal{G}) - npc(r, \mathcal{G})}{2 \cdot (npi(r, \mathcal{G}) + npc(r, \mathcal{G}))} + 0.5$$

Since the real-world KGs are often highly incomplete, it might be reasonable to put more weight on predictions in complete parts. This is done via combining any existing association rule measure rm, e.g., standard confidence or conviction, with the directional metric, using a certain dedicated weighting factor $\gamma \in [0...1]$. Formally,

$$weighted\_dm(r, \mathcal{G}) = \gamma \cdot rm(r, \mathcal{G}) + (1 - \gamma) \cdot dm(r, \mathcal{G})$$

Example 18. We have $dm(r'_1, \mathcal{G}') \approx 0.33$ and $dm(r'_2, \mathcal{G}') = 0.8$. With weighted directional metric using standard confidence ($rm = conf$), for $\gamma = 0.5$, we get $weighted\_dm(r'_1, \mathcal{G}') \approx 0.29$ and $weighted\_dm(r'_2, \mathcal{G}') \approx 0.48$.   □

Optimized Rule Evaluation. Most of state-of-the-art rule mining systems parallelize only the rule construction phase, while the quality of the rules are determined in a single thread. Huge size of KGs such as YAGO or Freebase prohibits their storage on a single machine. Therefore, the rule quality estimation is usually delegated to some third-party database management system. In practice this might not always be effective.

Unlike other state-of-the-art rule mining systems, the algorithm of *ontological pathfinding* (OP) [6] focuses on the efficient examination of the rule's quality. After candidate rules are constructed relying on a variation of [65], their quality is determined via a sequences of parallelization and optimization methods including KG partitioning, joining and pruning strategies.

5   Nonmonotonic Rule Learning

So far we have considered approaches for constructing Horn rules. However, these are not sufficiently expressive for representing incomplete human knowledge, and they are inadequate for capturing exceptions. Thus, Horn rules extracted from the existing KGs can potentially predict erroneous facts as shown in Sect. 1.

In this section, we provide an overview of approaches for learning nonmonotonic rules from large and incomplete KGs. First, we describe a method that relies on cross-talk among the extracted rules to guess their exceptions [29,76]. Then we present an approach that exploits embedding-based methods for knowledge graph completion during rule construction [34].

5.1   Revision-Based Method

Exception handling has been traditionally faced in ILP by learning nonmonotonic logic programs [8,35,41,64,66] (see Sect. 3). However, most of the existing methods assume that the data from which the rules are induced is complete.

In [29], a revision-based approach for extracting exception-enriched (i.e., nonmonotonic) rules from incomplete KGs has been proposed. It pre-processes a KG by projecting all binary facts into unary ones applying a form of propositionalization technique [39] (e.g., *livesIn*(*brad*, *berlin*) could be translated to *livesInBerlin*(*brad*) or *livesInCapital*(*brad*) with obvious loss of information) and adapts data mining methods designed for transaction data to extract Horn rules, which are then augmented with negated atoms. In [76], the approach has been extended to KGs in their original relational form. We now briefly summarize the ideas of [29,76].

In these works, the KG completion problem from Definition 9 is treated as a *theory revision* task, where, given a KG and a set of (previously learned) Horn rules, the goal is to revise this set by adding exceptions, such that the obtained rules have higher predictive accuracy than the original ones.

Since normally, the ideal graph $\mathcal{G}^i$ is not available, in order to estimate the quality of a revised ruleset, two generic quality functions q_{rm} and $q_{conflict}$ are devised, which both take as input a ruleset R and a KG $\mathcal{G}$ and output a real value, reflecting the suitability of R for data prediction. More specifically,

$$q_{rm}(R,\mathcal{G}) = \frac{\sum_{r\in R} rm(r,\mathcal{G})}{|R|}, \tag{4}$$

where rm is some standard association rule measure [3]. Conversely, $q_{conflict}$ estimates the number of conflicting predictions that the rules in R generate. To measure $q_{conflict}$ for R, an extended set of rules R^{aux} is created, which contains every revised rule in R together with its auxiliary version. For a rule $r : h(X,Y) \leftarrow B, not\ E$ in R, its auxiliary version r^{aux} is constructed by: (i) transforming r into a Horn rule by removing not from negated body atoms, and (ii) replacing the head predicate h of r with a newly introduced predicate not_h which intuitively should contain instances which are not in h.

Example 19. The auxiliary rule r_2^{aux} for the rule r_2 from Sect. 1 is as follows $r_2^{aux} : not\_livesIn(Y,Z) \leftarrow marriedTo(X,Y), livesIn(X,Z), researcher(Y)$, and it informally reflects that married people among whom one is a researcher often do not live together. For $r_3 : livesIn(X,Y) \leftarrow bornIn(X,Y), not\ immigrant(X)$ we similarly have $r_3^{aux} : not\_livesIn(X,Y) \leftarrow bornIn(X,Y), immigrant(X)$. If $R = \{r_1, r_2\}$ then $R^{aux} = \{r_1, r_1^{aux}, r_2, r_2^{aux}\}$.

Intuitively, $q_{conflict}(R,\mathcal{G})$ estimates the portion of conflicting predictions $livesIn(s,o), not\_livesIn(s,o)$ made by the rules in R^{aux}. The hypothesis of [29,76] is that for a set R of good rules with reasonable exceptions, the number of conflicting predictions produced by R^{aux} is small. □

Formally, based on statistics of both rules r and r^{aux} in a set of exception-enriched rules R, the measure $q_{conflict}$ is defined as

$$q_{conflict}(R,\mathcal{G}) = \sum_{p\in pred(R)} \frac{|c\,|\,p(c), not\_p(c) \in \mathcal{G}_{R^{aux}}|}{|c\,|\,not\_p(c) \in \mathcal{G}_{R^{aux}}|} \tag{5}$$

where $pred(R)$ stores predicates appearing in R.

Formally, the problem targeted in [29,76] is formulated as follows:

Definition 12 (Quality-based Horn Theory Revision). *Given a KG $\mathcal{G}$, a set of nonground Horn rules R_H mined from $\mathcal{G}$, and a quality function rm, find a set of rules R_{NM} obtained by adding negated atoms to $body(r)$ for some $r\in R_H$ s.t. (i) $q_{rm}(R_{NM},\mathcal{G})$ is maximal, and (ii) $q_{conflict}(R_{NM},\mathcal{G})$ is minimal.*

Given the huge size of KGs, finding the best possible revision is infeasible in practice. Thus, in [29,76] a heuristics-based method that computes an approximate solution to the above problem has been proposed (see Fig. 3 for overview). It combines standard relational association rule mining techniques with a FOIL-like supervised learning algorithm [57] for computing exceptions. We now briefly discuss the steps of this approach.

Step 1. After mining Horn rules using an off-the-shelf algorithm (e.g., [30]), one computes for each rule the *normal* and *abnormal substitutions*, defined as

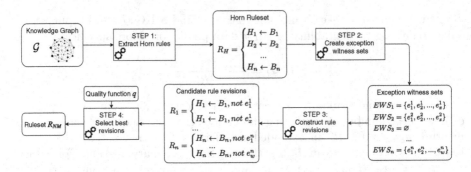

Fig. 3. Rule revision method for nonmonotonic rule learning [29,76].

Definition 13 (r-(Ab)Normal Substitutions). *Let $\mathcal{G}$ be a KG, r a Horn rule mined from $\mathcal{G}$, and let $\mathbf{V}$ be a set of variables occurring in r. Then*

- *$NS(r, \mathcal{G}) = \{\theta \mid head(r)\theta, body(r)\theta \subseteq \mathcal{G}\}$ is an r-normal set of substitutions;*
- *$ABS(r, \mathcal{G}) = \{\theta' \mid body(r)\theta' \subseteq \mathcal{G}, head(r)\theta' \not\subseteq \mathcal{G}\}$ is an r-abnormal set of substitutions, where $\theta, \theta' : \mathbf{V} \to \mathcal{C}$.*

Example 20. For $\mathcal{G}$ from Fig. 1 and r_1, we have $NS(r_1, \mathcal{G}) = \{\theta_1, \theta_2, \theta_3\}$, where $\theta_1 = \{X/brad, Y/ann, Z/berlin\}$, $\theta_2 = \{X/john, Y/kate, Z/chicago\}$ and $\theta_3 = \{X/sui, Y/li, Z/beijing\}$ respectively. Besides, among substitutions in $ABS(r_1, \mathcal{G})$, we have $\theta_4 = \{X/mat, Y/lucy, Z/amsterdam\}$, yet there are others. □

Step 2. Intuitively, if the given data was complete, then the r-normal and r-abnormal substitutions would exactly correspond to instances for which the rule r holds (respectively does not hold) in the real world. Since the KG is potentially incomplete, this is no longer the case and some r-abnormal instances might in fact be classified as such due to data incompleteness. In order to distinguish the "wrongly" and "correctly" classified instances in the r-abnormal set, one constructs *exception witness sets (EWS)*, which are defined as follows:

Definition 14 (Exception Witness Set (EWS)). *Let $\mathcal{G}$ be a KG, let r be a rule mined from it, let $\mathbf{V}$ be a set of variables occurring in r and $\mathbf{X} \subseteq \mathbf{V}$. Exception witness set for r w.r.t. $\mathcal{G}$ and $\mathbf{X}$ is a maximal set of predicates $EWS(r, \mathcal{G}, \mathbf{X}) = \{e_1, \ldots, e_k\}$, s.t.*

- *$e_i(\mathbf{X}\theta_j) \in \mathcal{G}$ for some $\theta_j \in ABS(r, \mathcal{G})$, $1 \le i \le k$ and*
- *$e_1(\mathbf{X}\theta'), \ldots, e_k(\mathbf{X}\theta') \notin \mathcal{G}$ for all $\theta' \in NS(r, \mathcal{G})$.*

Example 21. For $\mathcal{G}$ in Fig. 1 and rule r_1, we observe that $EWS(r, \mathcal{G}, Y) = \{researcher\}$ and $EWS(r, \mathcal{G}, X) = \{artist\}$. If *brad* with *ann* and *john* with *kate* did not live in metropolitan cities, then $EWS(r, \mathcal{G}, Z) = \{metropolitan\}$. □

In general, there are exponentially many possible EWSs to construct. Combinations of exception candidates could be an explanation for some missing KG edges, so the search space of solutions to the above problem is large. In [76] a restriction to a single atom as a final exception has been posed; extending exceptions to arbitrary combinations of atoms is left for future research.

Steps 3 and 4. After EWSs are computed for all rules in R_H, they are used to create potential revisions (Step 3), i.e., from every $e_j \in EWS(r_i, \mathcal{G})$ a revision r_i^j of r_i is constructed by adding a negated atom over e_j to the body of r_i. Finally, a concrete revision for every rule is determined, which constitutes a solution to the above problem (Step 4). To find such globally best ruleset revision R_{NM} many candidate combinations have to be checked, which due to the large size of $\mathcal{G}$ and EWSs might be too expensive. Thus, instead, R_{NM} is incrementally built by considering every $r_i \in R_H$ and choosing the locally best revision r_i^j for it.

In order to select r_i^j, four special ranking functions are introduced: a naive one and three more advanced functions, which exploit the concept of *partial materialization* (**PM**). Intuitively, the idea behind it is to rank candidate revisions not based on $\mathcal{G}$, but rather on its extension with predictions produced by other (selectively chosen) rules (grouped into a set R'), thus ensuring a cross-talk among the rules. We now describe the ranking functions in more details.

- **Naive** ranker is the most straightforward ranking function. It prefers the revision r_i^j with the highest value of $rm(r_i^j, \mathcal{G})$ among all revisions of r_i.
- **PM** ranking function prefers r_i^j with the highest value of

$$\frac{rm(r_i^j, \mathcal{G}_{R'}) + rm(r_i^{j\,aux}, \mathcal{G}_{R'})}{2} \qquad (6)$$

  where R' is the set of rules r_k', which are rules from $R_H \backslash r_i$ with all exceptions from $EWS(r_k, \mathcal{G})$ incorporated at once. Informally, $\mathcal{G}_{R'}$ contains only facts that can be safely predicted by the rules from $R_H \backslash r_i$, i.e., there is no evident reason (candidate exceptions) to neglect their predictions.
- **OPM** is similar to **PM**, but the selected ruleset R' contains only those rules whose Horn version appears above the considered rule r_i in the ruleset R_H, ordered (**O**) based on some chosen measure (e.g., the same as rm).
- **OWPM** is the most advanced ranking function. It differs from **OPM** in that the predicted facts in $\mathcal{G}_{R'} \backslash \mathcal{G}$ inherit weights (**W**) from the rules that produced them, and facts in $\mathcal{G}$ get the highest weight. These weights are taken into account when computing the value of (6). If the same fact is derived by multiple rules, the highest weight is stored. To avoid propagating uncertainty through rule chaining when computing weighted partial materialization of $\mathcal{G}$ predicted facts (i.e., derived by applying rules from R') are kept separately from the explicit facts (i.e., those in $\mathcal{G}$), and new facts are inferred using only $\mathcal{G}$.

For reasoning over weighted facts existing probabilistic deductive tools such as ProbLog [14, 26] can be used, but their exploitation is left for future work.

$r_{e1} : writtenBy(X, Z) \leftarrow hasPredecessor(X, Y), writtenBy(Y, Z), \textbf{not } american\_film(X)$
$r_{e2} : actedIn(X, Z) \leftarrow isMarriedTo(X, Y), directed(Y, Z), \textbf{not } silent\_film\_actor(X)$
$r_{e3} : isPoliticianOf(X, Z) \leftarrow hasChild(X, Y), isPoliticianOf(Y, Z), \textbf{not } vicepresidentOfMexico(X)$

Fig. 4. Examples of the mined rules [76]

Example Rules. Figure 4 shows examples of rules obtained in [76] by the described method. For instance, r_{e1} extracted from IMDB states that movie plot writers stay the same throughout the sequel unless a movie is American, while r_{e2} reflects that spouses of movie directors often appear on the cast with the exception of actors of old silent movies. Finally, the rule r_{e3} learned from YAGO says that ancestors of politicians are also politicians in the same country with the exception of Mexican vice-presidents.

5.2   Method Guided by Embedding Models

In the work [34] introducing RuLES framework, an alternative method for non-monotonic rule induction has been proposed, which first constructs an approximation of an ideal graph $\mathcal{G}^i$ and then learns rules by relying on it. For building such approximation, representations (i.e. embeddings) of entities and relations are learned from a given KG possibly enriched with additional information sources (e.g., text). We refer the reader to [78] for an overview of KG embedding models.

The RuLES approach [34] establishes a framework to benefit from the advantages of these models by iteratively inducing rules from a KG and collecting statistics about them from a precomputed embedding model to prune unpromising rule candidates.

Let $\mathcal{G}$ be a KG over the signature $\Sigma_{\mathcal{G}} = (\mathcal{R}, \mathcal{C})$. A *probabilistic KG* is a pair $\mathcal{P} = (\mathcal{G}, f)$, where $f : \mathcal{R} \times \mathcal{C} \times \mathcal{C} \to [0, 1]$ is a probability function over the facts over $\Sigma_{\mathcal{G}}$ such that for each atom $a \in \mathcal{G}$ it holds that $f(a) \geq ct$, where ct is a correctness threshold.

The goal of RuLES is to learn rules that do not only describe the available graph $\mathcal{G}$ well, but also predict highly probable facts based on the function f which relies on embeddings of the KG. For that, it utilizes a hybrid rule quality function:

$$\mu(r, \mathcal{P}) = (1 - \lambda) \times \mu_1(r, \mathcal{G}) + \lambda \times \mu_2(\mathcal{G}_r, \mathcal{P}).$$

where λ is a weight coefficient and μ_1 is any classical quality measure of r over $\mathcal{G}$ such that $\mu_1 : (r, \mathcal{G}) \mapsto \alpha \in [0, 1]$ (e.g., *standard confidence* or *PCA confidence* [30]). μ_2 measures the quality of $\mathcal{G}_r$ (*i.e.*, the extension of $\mathcal{G}$ resulting from executing the rule r) based on $\mathcal{P}$ given that $\mu_2 : (\mathcal{G}_r, \mathcal{P}) \mapsto \alpha \in [0, 1]$. To this end, μ_2 is defined as the average probability of the newly predicted facts in $\mathcal{G}_r$:

$$\mu_2(\mathcal{G}_r, \mathcal{P}) = \frac{\Sigma_{a \in \mathcal{G}_r \setminus \mathcal{G}} f(a)}{|\mathcal{G}_r \setminus \mathcal{G}|}.$$

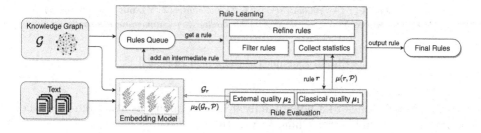

Fig. 5. Embedding-based method for nonmonotonic rule learning.

RuLES takes as input a KG, possibly a text corpus, and a set of user-specified parameters that are used to terminate rule construction. These parameters include an embedding weight λ, a minimum threshold for μ_1, a minimum rule support *r-supp* and other *rule-related* parameters such as a maximum number of positive and negative atoms allowed in $body(r)$. As illustrated in the overview of the RuLES system in Fig. 5, the KG and text corpus are used to train the embedding model that in turn is utilized to construct the probabilistic function f. The *Rule Learning* component computes the rules similar to [30] in an iterative fashion by applying refinement operators, starting from the head, by adding atoms to its body one after another until at least one of the termination criteria (that depend on f) is met. The set of refinement operators from [30] is extended by the following two to support negations in rule bodies:

- *add an exception instantiated atom:* add a binary negated atom with one of its arguments being a constant, and the other one being a shared variable.
- *add an exception closing atom:* add a binary negated atom to the rule with both of its arguments being shared variables.

During the construction of a rule r, the quality $\mu(r)$ is computed by the *Rule Evaluation* component, based on which a decision about the next action is made. Finally as an output, the RuLES system produces a set of nonmonotonic rules suitable for KG completion.

Note that the exploitation of the embedding feedback helps to distinguish exceptions from noise. Consider the rule r_1 stating that married people live together. This rule can have several possible exceptions, e.g., either one of the spouses is a researcher or he/she works at a company, which has headquarter in the US. Whenever the rule is enriched with an exception, naturally, the support of its body decreases, i.e., the size of $\mathcal{G}_r$ goes down. Ideally, such negated atoms should be added to the body of the rule that the average quality of $\mathcal{G}_r$ increases, as this witnesses that the addition of negated atoms to the rule body reduces unlikely predictions.

Example Rules. Figure 6 presents examples of rules learned by the RuLES system from the Wikidata KG. The first rule r_{e4} says that a person is a citizen of the country where his alma mater is located, unless it is a research institution, since most researchers in universities are foreigners. Additionally, r_{e5} encodes

$r_{e4} : nationality(X, Y) \leftarrow graduatedFrom(X, Z), inCountry(Z, Y),$ **not** $researchUni(Z)$

$r_{e5} : nobleFamily(X, Y) \leftarrow spouse(X, Z), nobleFamily(Z, Y),$ **not** $chineseDynasties(Y)$

Fig. 6. Examples of the mined rules [34]

that someone belongs to a noble family if his/her spouse is also from the same noble family, excluding the Chinese dynasties.

6   Discussion and Outlook

In this tutorial, we have presented a brief overview of the current techniques for rule induction from knowledge graphs and have demonstrated how reasoning over KGs using the learned rules can be exploited for KG completion.

While the problem of rule-based KG completion has recently gained a lot of attention, several promising research directions are still left unexplored.

Learning Other Rule Forms. The majority of available methods focus on extracting Horn or nonmonotonic rules, yet inducing rules of other, more complex, forms would be beneficial. These include disjunctive rules (e.g., *"Having a sibling implies having a sister or a brother"*, *"Korean speakers are normally either from South or North Korea"*) or rules with existential quantifiers in the head (e.g., *"Being a musicians in a band, implies playing some musical instrument"*). Several recent works on mining keys in KGs [40,74], detecting mandatory relations [40] and learning SHACL constraints [56] are relevant, but they do not directly address the mentioned rule forms. A combination of techniques from relational rule learning [58] and propositionalization approaches [39] can be utilized for learning rules with existentials, yet it is unclear how to detect KG parts that are worth propositionalizing and combine the outputs of both methods.

Rules that reflect correlations between edge counts in KGs such as *"If a person has two siblings then his/her parents are likely to have 3 children"* have been studied in [75]. Inducing more general rules, encoding mathematical functions on edge counts (e.g., *"If a person has k siblings then his/her parents are likely to have k + 1 children"*) and other numerical rules is still an open problem, which is particularly challenging due to large search space of possible hypothesis.

Learning temporal rules or constraints such as *"A person cannot graduate from a university before being born"* is another promising future work direction. Deductive reasoning over temporal KGs has been recently considered in [5]; however, the inductive setting has not yet been studied in full details to the best of our knowledge. A framework for learning hard boolean constraints has been described in [62], but its extension to KGs and soft constraints is still missing.

Learning Rules from Probabilistic Data. The majority of existing rule learning approaches over knowledge graphs model KGs as sets of true facts thus totally ignoring possible inaccuracies. Since KGs are usually constructed using

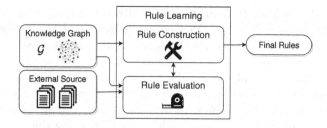

Fig. 7. Rule learning with external sources.

(semi-)automatic methods from noisy textual resources, obviously not all of the extracted facts should have the same weight. Learning rules from such noisy KGs by treating all facts equally might naturally lead to problematic rules, which when being applied may propagate faulty facts.

A recent ILP approach that accounts for noise has been proposed in [24]; but it relies on CWA and neglects weights on the facts. Learning rules from KGs treated as uncertain data sources has been considered in [10,59,63]; however, these works neglect negation, disjunction or existential variables in the head. Extending the techniques to more advanced rule forms is beneficial.

Rule Learning with External Sources. Another interesting research stream is to consider pieces of evidence from hybrid external sources while inducing rules from KGs (see Fig. 7). Similar to [22], where external functions are utilized during deductive reasoning, various heterogeneous information sources can be used to guide rule induction. These range from a human expert giving feedback about the correctness of a given rule (similar as done in [20] for pattern mining), to dedicated fact-checking engines (e.g., Defacto [70], FactChecker [52]) that given a fact such as *bornIn(einstein, ulm)* rely on Web documents to estimate its truthfulness.

Neural-Based Rule Learning. Utilizing embedding models for rule learning is a new research direction that has recently gained attention [80,81]. Most of the existing methods are purely statistics-based, i.e., they reduce the rule learning problem to algebraic operations on neural-embedding-based representations of a given KG. The approach [80] constructs rules by modeling relation composition as multiplication or addition of two relation embeddings. The authors of [81] propose a differentiable system for learning models defined by sets of first-order rules that exploits a connection between inference and sparse matrix multiplication [7]. These techniques pose strong restrictions on target rule patterns, which often prohibits learning interesting rules, e.g., non-chain-like or exception-aware ones. Combining neural methods with symbolic ones in a similar way as in [34] but also accounting for rich background knowledge in the form of logical theories is expected to be advantageous for obtaining surprising insights from the data.

Another ambitious direction is to mimic the active learning framework of Angluin *et al.* [2] for hypothesis discovery by issuing dedicated queries to possibly text-enhanced KG embedding models instead of a human expert.

Extracting Rules Jointly from KGs and Text. While modern KGs are rich in facts and typically rather clean, they contain a limited set of encyclopedic relations (e.g., *bornIn*, *marriedTo*). On the other hand, textual resources certainly cover a richer set of predicates (e.g., *gotAquintedWith*, *celebratedWedding*), but suffer from noise. A natural way to address the above issues is to combine text-based rule extraction relying on natural language processing (NLP) and textual entailment techniques [17,33,68] with inductive rule learning from KGs. This interesting research direction comes with many challenging due to the heterogeneity of the input sources.

References

1. Freebase: an open, shared database of the world's knowledge. http://www.freebase.com/
2. Angluin, D.: Queries and concept learning. Mach. Learn. **2**(4), 319–342 (1987)
3. Azevedo, P.J., Jorge, A.M.: Comparing rule measures for predictive association rules. In: Kok, J.N., Koronacki, J., Mantaras, R.L., Matwin, S., Mladenič, D., Skowron, A. (eds.) ECML 2007. LNCS (LNAI), vol. 4701, pp. 510–517. Springer, Heidelberg (2007). https://doi.org/10.1007/978-3-540-74958-5_47
4. Boytcheva, S.: Overview of inductive logic programming (ILP) systems (2007)
5. Chekol, M.W., Pirrò, G., Schoenfisch, J., Stuckenschmidt, H.: Marrying uncertainty and time in knowledge graphs. In: AAAI, pp. 88–94 (2017)
6. Chen, Y., Goldberg, S.L., Wang, D.Z., Johri, S.S.: Ontological pathfinding. In: SIGMOD, pp. 835–846. ACM (2016)
7. Cohen, W.W.: TensorLog: a differentiable deductive database. CoRR abs/1605.06523 (2016)
8. Corapi, D., Russo, A., Lupu, E.: Inductive logic programming as abductive search. In: ICLP, pp. 54–63 (2010)
9. Corapi, D., Russo, A., Lupu, E.: Inductive logic programming in answer set programming. In: Muggleton, S.H., Tamaddoni-Nezhad, A., Lisi, F.A. (eds.) ILP 2011. LNCS (LNAI), vol. 7207, pp. 91–97. Springer, Heidelberg (2012). https://doi.org/10.1007/978-3-642-31951-8_12
10. Corapi, D., Sykes, D., Inoue, K., Russo, A.: Probabilistic rule learning in nonmonotonic domains. In: Leite, J., Torroni, P., Ågotnes, T., Boella, G., van der Torre, L. (eds.) CLIMA 2011. LNCS (LNAI), vol. 6814, pp. 243–258. Springer, Heidelberg (2011). https://doi.org/10.1007/978-3-642-22359-4_17
11. d'Amato, C., Staab, S., Tettamanzi, A.G., Minh, T.D., Gandon, F.: Ontology enrichment by discovering multi-relational association rules from ontological knowledge bases. In: SAC, pp. 333–338 (2016)
12. Darari, F., Nutt, W., Pirrò, G., Razniewski, S.: Completeness statements about RDF data sources and their use for query answering. In: Alani, H., et al. (eds.) ISWC 2013. LNCS, vol. 8218, pp. 66–83. Springer, Heidelberg (2013). https://doi.org/10.1007/978-3-642-41335-3_5
13. De Raedt, L., Bruynooghe, M.: CLINT : a multi-strategy interactive concept-learner and theory revision system. In: Michalski, R., Tecuci, G. (eds.) Proceedings of the Multi-Strategy Learning Workshop, pp. 175–191 (1991)
14. De Raedt, L., Kimmig, A., Toivonen, H.: ProbLog: a probabilistic prolog and its application in link discovery. In: IJCAI, pp. 2468–2473 (2007)

15. Dehaspe, L., De Raedt, L.: Mining association rules in multiple relations. In: Lavrač, N., Džeroski, S. (eds.) ILP 1997. LNCS, vol. 1297, pp. 125–132. Springer, Heidelberg (1997). https://doi.org/10.1007/3540635149_40
16. Dong, X., et al.: Knowledge vault: a web-scale approach to probabilistic knowledge fusion. In: Proceedings of the 20th ACM SIGKDD International Conference on Knowledge Discovery and Data Mining, pp. 601–610. ACM (2014)
17. Dragoni, M., Villata, S., Rizzi, W., Governatori, G.: Combining NLP approaches for rule extraction from legal documents. In: 1st Workshop on MIning and REasoning with Legal Texts (MIREL) (2016)
18. Duc Tran, M., d'Amato, C., Nguyen, B.T., Tettamanzi, A.G.B.: Comparing rule evaluation metrics for the evolutionary discovery of multi-relational association rules in the semantic web. In: Castelli, M., Sekanina, L., Zhang, M., Cagnoni, S., García-Sánchez, P. (eds.) EuroGP 2018. LNCS, vol. 10781, pp. 289–305. Springer, Cham (2018). https://doi.org/10.1007/978-3-319-77553-1_18
19. Dzeroski, S., Lavrac, N.: Learning relations from noisy examples: An empirical comparison of LINUS and FOIL. In: ML (1991)
20. Dzyuba, V., van Leeuwen, M.: Learning what matters – sampling interesting patterns. In: Kim, J., Shim, K., Cao, L., Lee, J.-G., Lin, X., Moon, Y.-S. (eds.) PAKDD 2017. LNCS (LNAI), vol. 10234, pp. 534–546. Springer, Cham (2017). https://doi.org/10.1007/978-3-319-57454-7_42
21. Eiter, T., Ianni, G., Krennwallner, T.: Answer set programming: a primer. In: Tessaris, S. (ed.) Reasoning Web 2009. LNCS, vol. 5689, pp. 40–110. Springer, Heidelberg (2009). https://doi.org/10.1007/978-3-642-03754-2_2
22. Eiter, T., Kaminski, T., Redl, C., Schüller, P., Weinzierl, A.: Answer set programming with external source access. In: Ianni, G., et al. (eds.) Reasoning Web 2017. LNCS, vol. 10370, pp. 204–275. Springer, Cham (2017). https://doi.org/10.1007/978-3-319-61033-7_7
23. Elkan, C., Noto, K.: Learning classifiers from only positive and unlabeled data. In: ACM SIGKDD International Conference on Knowledge Discovery and Data Mining, pp. 213–220 (2008)
24. Evans, R., Grefenstette, E.: Learning explanatory rules from noisy data. J. Artif. Intell. Res. **61**, 1–64 (2018)
25. Faber, W., Pfeifer, G., Leone, N.: Semantics and complexity of recursive aggregates in answer set programming. Artif. Intell. **175**(1), 278–298 (2011)
26. Fierens, D., et al.: Inference and learning in probabilistic logic programs using weighted boolean formulas. TPLP **15**(3), 358–401 (2015)
27. Fürnkranz, J., Gamberger, D., Lavrac, N.: Foundations of Rule Learning. Cognitive Technologies. Springer, Heidelberg (2012). https://doi.org/10.1007/978-3-540-75197-7
28. Fürnkranz, J., Kliegr, T.: A brief overview of rule learning. In: Bassiliades, N., Gottlob, G., Sadri, F., Paschke, A., Roman, D. (eds.) RuleML 2015. LNCS, vol. 9202, pp. 54–69. Springer, Cham (2015). https://doi.org/10.1007/978-3-319-21542-6_4
29. Gad-Elrab, M.H., Stepanova, D., Urbani, J., Weikum, G.: Exception-enriched rule learning from knowledge graphs. In: Groth, P., et al. (eds.) ISWC 2016. LNCS, vol. 9981, pp. 234–251. Springer, Cham (2016). https://doi.org/10.1007/978-3-319-46523-4_15
30. Galarraga, L., Teflioudi, C., Hose, K., Suchanek, F.M.: Fast rule mining in ontological knowledge bases with AMIE+. VLDB **24**, 707–730 (2015)
31. Gelfond, M., Lifschitz, V.: The stable model semantics for logic programming. In: Proceedings of the 5th International Conference and Symposium on Logic Programming, ICLP 1988, pp. 1070–1080 (1988)

32. Goethals, B., Van den Bussche, J.: Relational association rules: getting WARMER. In: Hand, D.J., Adams, N.M., Bolton, R.J. (eds.) Pattern Detection and Discovery. LNCS (LNAI), vol. 2447, pp. 125–139. Springer, Heidelberg (2002). https://doi.org/10.1007/3-540-45728-3_10

33. Gordon, J., Schubert, L.K.: Discovering commonsense entailment rules implicit in sentences. In: TextInfer Workshop on Textual Entailment, TIWTE 2011, pp. 59–63 (2011)

34. Ho, V.T., Stepanova, D., Gad-Elrab, M.H., Kharlamov, E., Weikum, G.: Rule learning from knowledge graphs guided by embedding models. In: ISWC 2018 (2018, in print)

35. Inoue, K., Kudoh, Y.: Learning extended logic programs. In: IJCAI, pp. 176–181. Morgan Kaufmann (1997)

36. Józefowska, J., Lawrynowicz, A., Lukaszewski, T.: The role of semantics in mining frequent patterns from knowledge bases in description logics with rules. TPLP 10(3), 251–289 (2010)

37. Katzouris, N., Artikis, A., Paliouras, G.: Incremental learning of event definitions with inductive logic programming. Mach. Learn. 100(2–3), 555–585 (2015)

38. Klyne, G., Carroll, J.J.: Resource description framework (RDF): concepts and abstract syntax. W3C Recommendation (2004)

39. Krogel, M.-A., Rawles, S., Železný, F., Flach, P.A., Lavrač, N., Wrobel, S.: Comparative evaluation of approaches to propositionalization. In: Horváth, T., Yamamoto, A. (eds.) ILP 2003. LNCS (LNAI), vol. 2835, pp. 197–214. Springer, Heidelberg (2003). https://doi.org/10.1007/978-3-540-39917-9_14

40. Lajus, J., Suchanek, F.M.: Are all people married?: determining obligatory attributes in knowledge bases. In: WWW, pp. 1115–1124. ACM (2018)

41. Law, M., Russo, A., Broda, K.: The ILASP system for learning answer set programs (2015). https://www.doc.ic.ac.uk/~ml1909/ILASP

42. Lehmann, J., et al.: DBpedia - a large-scale, multilingual knowledge base extracted from Wikipedia. Semant. Web 6, 167–195 (2015)

43. Lisi, F.A.: Inductive logic programming in databases: from datalog to DL+log. TPLP 10(3), 331–359 (2010)

44. Miller, G.A.: WordNet: a lexical database for English. Commun. ACM 38(11), 39–41 (1995)

45. Mirza, P., Razniewski, S., Darari, F., Weikum, G.: Cardinal virtues: extracting relation cardinalities from text. In: ACL (2017)

46. Mirza, P., Razniewski, S., Nutt, W.: Expanding wikidata's parenthood information by 178%, or how to mine relation cardinality information. In: ISWC 2016 Posters and Demos (2016)

47. Mitchell, T., et al.: Never-ending learning. In: AAAI, pp. 2302–2310 (2015)

48. Morik, K.: Balanced cooperative modeling. Mach. Learn. 11(2), 217–235 (1993)

49. Muggleton, S.: Inductive logic programming. New Gener. Comput. 8(4), 295–318 (1991)

50. Muggleton, S., Buntine, W.L.: Machine invention of first order predicates by inverting resolution. In: International Conference on Machine Learning, pp. 339–352 (1988)

51. Muggleton, S., Feng, C.: Efficient induction of logic programs. In: Algorithmic Learning Theory Workshop, pp. 368–381 (1990)

52. Nakashole, N., Mitchell, T.M.: Language-aware truth assessment of fact candidates. In: Proceedings of the 52nd Annual Meeting of the Association for Computational Linguistics, vol. 1: Long Papers, pp. 1009–1019 (2014)

53. Nickel, M., Murphy, K., Tresp, V., Gabrilovich, E.: A review of relational machine learning for knowledge graphs: from multi-relational link prediction to automated knowledge graph construction. CoRR (2015)
54. Nickel, M., Murphy, K., Tresp, V., Gabrilovich, E.: A review of relational machine learning for knowledge graphs. IEEE **104**(1), 11–33 (2016)
55. Paulheim, H.: Knowledge graph refinement: a survey of approaches and evaluation methods. Semant. Web **8**(3), 489–508 (2017)
56. Paulheim, H.: Learning SHACL constraints for validation of relation assertions in knowledge graphs. In: ESWC (2018, to appear)
57. Quinlan, J.R.: Learning logical definitions from relations. Mach. Learn. **5**, 239–266 (1990)
58. Raedt, L.D.: Logical and Relational Learning. Cognitive Technologies. Springer, Heidelberg (2008). https://doi.org/10.1007/978-3-540-68856-3
59. Raedt, L.D., Dries, A., Thon, I., den Broeck, G.V., Verbeke, M.: Inducing probabilistic relational rules from probabilistic examples. In: IJCAI, pp. 1835–1843. AAAI Press (2015)
60. Raedt, L.D., Dzeroski, S.: First-order jk-clausal theories are PAC-learnable. Artif. Intell. **70**(1–2), 375–392 (1994)
61. Raedt, L.D., Lavrac, N., Dzeroski, S.: Multiple predicate learning. In: Proceedings of the 13th International Joint Conference on Artificial Intelligence, Chambéry, France, 28 August–3 September 1993, pp. 1037–1043 (1993)
62. Raedt, L.D., Passerini, A., Teso, S.: Learning constraints from examples. In: AAAI (2018)
63. De Raedt, L., Thon, I.: Probabilistic rule learning. In: Frasconi, P., Lisi, F.A. (eds.) ILP 2010. LNCS (LNAI), vol. 6489, pp. 47–58. Springer, Heidelberg (2011). https://doi.org/10.1007/978-3-642-21295-6_9
64. Ray, O.: Nonmonotonic abductive inductive learning. J. Appl. Log. **7**(3), 329–340 (2009). Special Issue: Abduction and Induction in Artificial Intelligence
65. Richards, B.L., Mooney, R.J.: Learning relations by pathfinding. In: Proceedings of the 10th National Conference on Artificial Intelligence, pp. 50–55 (1992)
66. Sakama, C.: Induction from answer sets in nonmonotonic logic programs. ACM Trans. Comput. Log. **6**(2), 203–231 (2005)
67. Sazonau, V., Sattler, U.: Mining hypotheses from data in OWL: advanced evaluation and complete construction. In: d'Amato, C. (ed.) ISWC 2017. LNCS, vol. 10587, pp. 577–593. Springer, Cham (2017). https://doi.org/10.1007/978-3-319-68288-4_34
68. Schoenmackers, S., Etzioni, O., Weld, D.S., Davis, J.: Learning first-order horn clauses from web text. In: EMNLP, pp. 1088–1098 (2010)
69. Shapiro, E.Y.: Algorithmic Program DeBugging. MIT Press, Cambridge (1983)
70. Speck, R., Esteves, D., Lehmann, J., Ngonga Ngomo, A.C.: Defacto - a multilingual fact validation interface. In: ISWC (2015)
71. Srinivasan, A.: The aleph manual. http://www.cs.ox.ac.uk/activities/machlearn/Aleph/aleph.html
72. Suchanek, F.M., Kasneci, G., Weikum, G.: YAGO: a core of semantic knowledge. In: Proceedings of WWW, pp. 697–706 (2007)
73. Suchanek, F.M., Preda, N.: Semantic culturomics. VLDB **7**(12), 1215–1218 (2014)
74. Symeonidou, D., Galárraga, L., Pernelle, N., Saïs, F., Suchanek, F.: VICKEY: mining conditional keys on knowledge bases. In: d'Amato, C., et al. (eds.) ISWC 2017. LNCS, vol. 10587, pp. 661–677. Springer, Cham (2017). https://doi.org/10.1007/978-3-319-68288-4_39

75. Tanon, T.P., Stepanova, D., Razniewski, S., Mirza, P., Weikum, G.: Completeness-aware rule learning from knowledge graphs. In: ISWC, pp. 507–525 (2017)
76. Tran, H.D., Stepanova, D., Gad-Elrab, M.H., Lisi, F.A., Weikum, G.: Towards nonmonotonic relational learning from knowledge graphs. In: Cussens, J., Russo, A. (eds.) ILP 2016. LNCS (LNAI), vol. 10326, pp. 94–107. Springer, Cham (2017). https://doi.org/10.1007/978-3-319-63342-8_8
77. Vrandecic, D., Krötzsch, M.: Wikidata: a free collaborative knowledgebase. CACM **57**(10), 78–85 (2014)
78. Wang, Q., Mao, Z., Wang, B., Guo, L.: Knowledge graph embedding: a survey of approaches and applications, pp. 2724–2743 (2017)
79. Wang, Z., Li, J.: RDF2Rules: learning rules from RDF knowledge bases by mining frequent predicate cycles. CoRR abs/1512.07734 (2015)
80. Yang, B., Yih, W., He, X., Gao, J., Deng, L.: Embedding entities and relations for learning and inference in knowledge bases. CoRR abs/1412.6575 (2014)
81. Yang, F., Yang, Z., Cohen, W.W.: Differentiable learning of logical rules for knowledge base reasoning. In: NIPS, pp. 2316–2325 (2017)
82. Zupanc, K., Davis, J.: Estimating rule quality for knowledge base completion with the relationship between coverage assumption. In: WWW, pp. 1073–1081 (2018)

Storing and Querying Semantic Data in the Cloud

Daniel Janke[1(✉)] and Steffen Staab[1,2]

[1] Institute for Web Science and Technologies, Universität Koblenz-Landau,
Koblenz, Germany
{danijank,staab}@uni-koblenz.de
[2] Web and Internet Science Group, University of Southampton, Southampton, UK
s.r.staab@soton.ac.uk,
http://west.uni-koblenz.de/, http://wais.ecs.soton.ac.uk/

Abstract. In the last years, huge RDF graphs with trillions of triples were created. To be able to process this huge amount of data, scalable RDF stores are used, in which graph data is distributed over compute and storage nodes for scaling efforts of query processing and memory needs. The main challenges to be investigated for the development of such RDF stores in the cloud are: (i) strategies for data placement over compute and storage nodes, (ii) strategies for distributed query processing, and (iii) strategies for handling failure of compute and storage nodes. In this manuscript, we give an overview of how these challenges are addressed by scalable RDF stores in the cloud.

1 Introduction

In the last years, huge RDF graphs with trillions of triples were created. For instance, the number of Schema.org-based facts that are extracted out of the Web have reached the size of three trillions [98]. Another example is the European Bioinformatics Institute (EMBL-EBI) that would like to convert its datasets into RDF resulting in a graph consisting of trillions of triples. To date no such scalable RDF store exists and the current EBI RDF Platform can handle only 10 billion triples [88].

In order to provide RDF stores that can scale to these huge graph sizes, researchers have started to develop RDF stores in the cloud, where graph data is distributed over compute and storage nodes for scaling efforts of query processing and memory needs. The main challenges to be investigated for such development are: (i) strategies for data placement over compute and storage nodes, (ii) strategies for distributed query processing, and (iii) strategies for handling failure of compute and storage nodes. In this manuscript, we want to give an overview of how these challenges have been addressed by scalable RDF stores in the cloud that have been developed in the last 15 years.

In Sect. 3, we give an overview of the different architectures of scalable RDF stores in the cloud. Basically there exist three types of architectures. The first

© Springer Nature Switzerland AG 2018
C. d'Amato and M. Theobald (Eds.): Reasoning Web 2018, LNCS 11078, pp. 173–222, 2018.
https://doi.org/10.1007/978-3-030-00338-8_7

type uses general cluster computing frameworks like Spark[1] or HBase[2] to perform queries on RDF graphs. The second type – so-called distributed RDF stores— splits the RDF graph into smaller parts that are then stored and queried on the compute and storage nodes. The last type are federated query processing systems. These systems do not have any influence on the data distribution over compute and storage nodes. Instead, they process queries over several RDF stores.

In case of cluster computing frameworks and distributed RDF stores, the RDF graph is distributed among several compute and storage nodes. The general procedure of data distribution strategies is to first split the graph into several not-necessarily disjoint triple sets. In relational and NoSQL databases this splitting is called sharding. Thereafter, the individual triple sets are assigned to compute and storage nodes. An overview of how these two steps are preformed by the existing RDF stores is given in Sect. 4.

Querying a distributed dataset is usually done by splitting the query into subqueries that can be answered locally without the need to exchange data over the network. The results of these subqueries are finally combined into the overall results of the query. In order to identify which parts of the queries can be answered locally on which compute nodes, an index is required that stores information about the data distribution. The different types of indices are described in Sect. 5. An overview of how distributed query processing is done by RDF stores in the cloud is given in Sect. 6.

Another challenge of scalable RDF stores in the cloud is that the failure of an individual compute or storage node should not lead to the failure of the complete RDF store. A brief overview of how this challenge is addressed is given in Sect. 7.

In order to evaluate how well RDF stores in the cloud solve the challenges in the cloud, their performance needs to be evaluated. In Sect. 8, we will present different methodologies how RDF stores in the cloud are evaluated.

Due to the huge amount of RDF stores in the cloud that were developed in the last 15 years, we will not present distributed solutions for the handling of RDF streams or reasoning. Interested readers are referred to [118]. Beside RDF stores in the cloud there also exist distributed graph databases like Sparksee[3] or Titan[4] as well as distributed graph processing frameworks like PGX.D [61] or PEGASUS [66]. They are not described in this manuscript since they have not been presented as part of an RDF store, yet. Furthermore, this manuscript gives an overview of how RDF stores in the cloud work. Readers interested in a performance comparison of RDF stores in the cloud are referred to, e.g., [49].

2   Preliminaries

RDF stores in the cloud have to deal with the two challenges how to distribute RDF graphs on several compute and storage nodes and how to retrieve data

---

[1] https://spark.apache.org/.

[2] https://hbase.apache.org/.

[3] http://www.sparsity-technologies.com/.

[4] http://titan.thinkaurelius.com/.

thereafter again. The following two sections formalize these challenges. The given definitions are the usual definitions found in the literature. This section was taken from [64].

2.1 Formalization of Graph Cover Strategies

In order to illustrate different graph cover strategies, we use Fig. 1 as our running example. The graph represents the knows relationship between two employees of the university institute WeST and one employee of the Leibniz institute GESIS. Additionally, the graph includes the ownership of the dog Bello. The terms r:, e:, w:, g:, and f: abbreviate IRI prefixes.

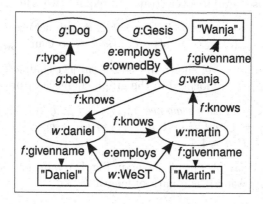

Fig. 1. The example graph describing the knows relationships between some employees of WeST and Gesis.

To formalize the data distribution challenge, we define RDF graphs like in [53]. Assume a signature $\sigma = (I, B, L)$, where I, B and L are the pairwise disjoint infinite sets of IRIs, blank nodes and literals, respectively. The union of these sets is abbreviated as *IBL*.

Definition 1. *The set of all possible RDF triples T for signature σ is defined by $T = (I \cup B) \times I \times IBL$. An RDF graph G or simply graph is defined as $G \subseteq T$. The set of all vertices contained in G is defined by $V_G = \{v | \exists s, p, o : (v, p, o) \in G \vee (s, p, v) \in G\}$.*

$(s, p, o) \in T$ is also called a triple with *subject s*, *property p* and *object o*. To simplify later definitions, the functions $\text{subj}(t)$, $\text{obj}(t)$ and $\text{prop}(t)$ return the subject, object or property of triple t, respectively. Likewise, we use $\text{subj}(T)$, $\text{obj}(T)$ and $\text{prop}(T)$ to refer to the set of subjects, objects and properties in the triple set T.

In the context of distributed RDF stores, the triples of a graph have to be assigned to different compute and storage nodes (in the following, we refer to them more briefly as *compute nodes*). The finite set of compute nodes is denoted as C in the rest of this paper.

Definition 2. *Let G denote an RDF graph. Then a* graph cover *is a function* cover: $G \to 2^C$, *that assigns each triple of a graph G to at least one compute node.*

Definition 3. *The function* chunk *returns the triples assigned to a specific compute node by a graph cover (*graph chunks*). It is defined as*

$$\text{chunk}_{\text{cover}}: C \to 2^G$$

$$\text{chunk}_{\text{cover}}(c) := \{t | c \in \text{cover}(t)\} .$$

This definition allows for triples being replicated on several compute nodes. If the graph chunks are pairwise disjoint the underlying graph cover is called a *graph partitioning*.[5]
Some frequently used graph cover strategies require two additional definitions.

Definition 4. *A* graph cover *of RDF graph G is* subject-complete, *if $\forall c \in C :$ $\forall (s, p, o) \in \text{chunk}_{\text{cover}}(c) : \forall (s, p', o') \in G : (s, p', o') \in \text{chunk}_{\text{cover}}(c)$.*

Example 1. The graph cover shown in Fig. 2 is subject-complete, since all triples with the same subject are located in the graph chunk of c_1 or c_2.

Definition 5. *A path P is a sequence $\langle t_0, t_1, \ldots, t_n \rangle$, if $\forall i \in [0, n] : t_i \in G \wedge \forall j \in [0, n] : j \neq i \Rightarrow t_j \neq t_i$ and $\forall i \in [1, n] : t_{i-1} = (s_{i-1}, p_{i-1}, s_i) \wedge t_i = (s_i, p_i, o_i)$. The* length *of path P is $n + 1$.*

Example 2. In the example Graph shown in Fig. 1, $\langle$(w:daniel, f:knows, w:martin), (w:martin, f:knows, g:wanja)$\rangle$ is a path of length 2.

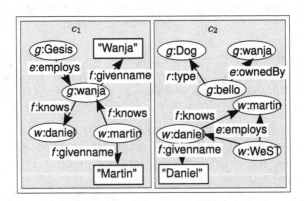

Fig. 2. An example graph cover of the example graph.

Several RDF stores in the cloud create subsets of an RDF graph that are distributed among the compute nodes. One frequently used graph subset is called

---

[5] In the context of relational or NoSQL databases, graph covers are called sharding and the graph chunks shards. In the literature, there exist definitions of sharding that allow for data replication whereas others do not allow it.

a molecule. A molecule is a set of triples that are reachable from its anchor vertex. This property of molecules aims for an efficient processing of star-shaped queries, in which the triple patterns are joined on the subject. Also path-shaped queries might be processed efficiently, if the length of the path is smaller than the diameter of the molecule.

Definition 6. *A subset $M_v \subseteq G$ of an RDF graph G is called* molecule[6] *of vertex v if*

1. *for all triples $t \in M_v$ there must exist a path $\langle t_0, t_1, \ldots, t_n, t \rangle$ such that $t_0 = (v, p_0, o_0)$ and $t_0, t_1, \ldots, t_n \in M_v$ and*
2. *if s is a subject of some triple in M_v, then $\forall (s, p, o) \in G : (s, p, o) \in M_v$.*

The vertex v is called anchor vertex[7].

Definition 7. *The* directed molecule diameter *of a molecule M_v is the longest shortest path between the anchor vertex v and all objects contained in triples of M_v.*

Example 3. The molecule with directed diameter 1 of the anchor vertex g:gesis in the example graph only contains the triple (g:Gesis, e:employs, g:wanja). A molecule with diameter 2 of the same anchor vertex would additionally contain (g:wanja, f:knows, w:daniel) and (g:wanja, f:givenname, "Wanja").

2.2   Formalization of Query Execution Strategies

We define the used SPARQL core as done in [16,107,110]. For this definition the infinite set of variables V that is disjoint from IBL is required. In order to distinguish the syntax of variables from other RDF terms, they are prefixed with ?. The syntax of SPARQL is defined as follows.

Definition 8. *A* triple pattern *is a member of the set $TP = (I \cup L \cup V) \times (I \cup V) \times (I \cup L \cup V)$.*

Definition 9. *A* basic graph pattern *(BGP) is a*

1. *triple pattern or*
2. *a conjunction $B_1.B_2$ of two BGPs B_1 and B_2.*

Definition 10. *A* SELECT *query is defined as* SELECT W WHERE $\{B\}$ *with $W \subseteq V$ and B a BGP.*

Example 4. The following SELECT query returns the names of all persons who are known by employees of WeST and who own the dog Bello. It contains a basic graph pattern that concatenates four triple patterns. In the following examples ?v1 <f:knows> ?v2 is abbreviated as tp_1, ?v2 <f:givenname> ?v3 as tp_2 and so on. All following examples in this section will refer to this query.

---

[6] We adapted the definition of an RDF molecule in [38] to allow for paths with a length ≥ 1.
[7] The term anchor vertex was taken from [79].

```
SELECT ?v3 WHERE {
    ?v1 <f:knows>  ?v2.
    ?v2 <f:givenname>  ?v3.
    <w:WeST> <e:employs> ?v1.
    <gs:bello> <e:ownedBy> ?v2
}
```

Before the semantics of a SPARQL query can be defined, some additional definitions are required. In the following Q represents the set of all SPARQL queries.

Definition 11. *The function* var $: Q \rightarrow V$ *returns the set of variables occurring in a SPARQL query. It is defined as:*

1. var(tp) *is the set of variables occurring in triple pattern tp.*
2. var$(B_1.B_2) := $ var$(B_1) \cup$ var(B_2) *for the conjunction of the BGPs B_1 and B_2.*
3. var$($SELECT W WHERE $\{B\}) := W \cap$ var(B) *for $W \subseteq V$ and B a BGP.*

Definition 12. *A* variable binding *is a partial function* $\mu : V \nrightarrow IBL$. *The* set *of all variable bindings is* $\mathcal{O}$.

The abbreviated notation $\mu(t)$ with $t \in TP$ means that the variables in t are substituted according to μ.

Example 5. The following three partial functions are variable bindings, that assign values to some variables. μ_1 would be an intermediate result produced by the first triple pattern of the example query in Example 4 whereas μ_2 and μ_3 would be produced by the second triple pattern.

$$\mu_1 = \{(?v_1, \mathsf{w{:}martin}), (?v_2, \mathsf{g{:}wanja})\}$$
$$\mu_2 = \{(?v_2, \mathsf{g{:}wanja}), (?v_3, "\mathsf{Wanja}")\}$$
$$\mu_3 = \{(?v_2, \mathsf{w{:}martin}), (?v_3, "\mathsf{Martin}")\}$$

Definition 13. *Two variable bindings μ_i and μ_j are* compatible, *denoted by* $\mu_i \sim \mu_j$, *if* $\forall ?x \in \mathrm{dom}(\mu_i) \cap \mathrm{dom}(\mu_j) : \mu_i(?x) = \mu_j(?x)$.[8]

Example 6. The variable bindings μ_1 and μ_2 from Example 5 are compatible since in both variable bindings g:wanja is assigned to $?v_2$ which is the only variable occurring in the domains of both variable bindings. μ_1 and μ_3 as well as μ_2 and μ_3 are not compatible because they assign different values to common variables.

Definition 14. *The* join *of two sets of variable bindings Ω_1 and Ω_2 is defined as* $\Omega_1 \bowtie \Omega_2 = \{\mu_1 \cup \mu_2 | \mu_1 \in \Omega_1 \wedge \mu_2 \in \Omega_2 \wedge \mu_1 \sim \mu_2\}$.
The variables contained in $\mathrm{dom}(\mu_1) \cap \mathrm{dom}(\mu_2)$ *are called* join variables.

---

[8] $\mathrm{dom}(\mu)$ refers to the set of variables of this binding.

Example 7. The join of the two variable bindings sets $\{\mu_1\}$ and $\{\mu_2, \mu_3\}$ from Example 5 produces a result set only containing the variable binding $\{(?v_1, \text{w:martin}), (?v_2, \text{g:wanja}), (?v_3, "\text{Wanja}")\}$ because only μ_1 and μ_2 are compatible.

[16,107] define the semantics of a SPARQL query as follows:

Definition 15. *The evaluation of a SPARQL query Q over an RDF Graph G, denoted by $[\![Q]\!]_G$, is defined recursively as follows:*

1. *If $tp \in TP$ then $[\![tp]\!]_G = \{\mu | \text{dom}(\mu) = \text{var}(tp) \wedge \mu(tp) \in G\}$.*
2. *If B_1 and B_2 are BGPs, then $[\![B_1.B_2]\!]_G = [\![B_1]\!]_G \bowtie [\![B_2]\!]_G$.*
3. *If $W \subseteq V$ and B is a BGP, then $[\![\text{SELECT } W \text{ WHERE } \{B\}]\!]_G = \text{project}(W, [\![B]\!]_G) = \{\mu_{|W} | \mu \in [\![B]\!]_G\}$.[9]*

The execution of a query requires the translation of a SPARQL query into a query execution tree. This tree defines the individual operations and their execution sequence. Thereby, each node of the query execution tree consists of three components: (i) the name of the operation to be executed, (ii) the set of variables that are bound in the resulting variable bindings and (iii) the set of child operations.

Definition 16. *Let L_{node} be the set of node labels and $\Upsilon = L_{node} \times 2^V \times 2^\Upsilon$ the set of all query execution trees, then a query execution tree of a query Q, denoted as $\langle\!\langle Q \rangle\!\rangle$, is defined recursively as follows:*

1. *If $tp \in TP$ then $\langle\!\langle tp \rangle\!\rangle = (tp, \text{var}(tp), \varnothing)$.*
2. *If B_1 and B_2 are BGPs, then $\langle\!\langle B_1.B_2 \rangle\!\rangle = (\text{join}, \text{var}(B_1) \cup \text{var}(B_2), \{\langle\!\langle B_1 \rangle\!\rangle, \langle\!\langle B_2 \rangle\!\rangle\})$.*
3. *If $W \subseteq V$ and B is a BGP, then $\langle\!\langle \text{SELECT } W \text{ WHERE } \{B\} \rangle\!\rangle = (\text{project}, W, \langle\!\langle B \rangle\!\rangle)$.*

Example 8. Figure 3 shows a graphical representation of one query execution tree for the example query from Example 4. The following query execution tree represents the first join in its mathematical representation. It has the two child trees $(tp_1, \{?v_1, ?v_2\}, \varnothing)$ and $(tp_2, \{?v_2, ?v_3\}, \varnothing)$.

$$
\begin{aligned}
(\text{join}, &\{?v_1, ?v_2, ?v_3\}, \{ \\
&(tp_1, \{?v_1, ?v_2\}, \varnothing), \\
&(tp_2, \{?v_2, ?v_3\}, \varnothing) \\
\})&
\end{aligned}
$$

---

[9] $\mu_{|W}$ means that the domain of μ is restricted to the variables in W.

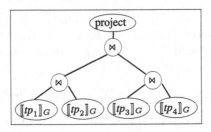

Fig. 3. Bushy query execution tree for the query from Example 4.

3   Architectures

The RDF stores in the cloud can be categorized into three groups that characterize their architecture. The first type of RDF stores in the cloud make use of cloud computing frameworks (see Sect. 4). These frameworks hide the complexity of distributed systems from the developers. This reduced developing complexity comes with the cost of limited influence on, e.g., the data placement. To overcome these limitation, developers of distributed RDF stores have to address the challenges of data placement, distributed query processing and fault tolerance on their own (see Sect. 3.2). In contrast to this, federated RDF stores aim to query data from several RDF stores that manage the stored data on their own (see Sect. 3.3). One application scenario would be querying several remote SPARQL endpoints that can be found in the linked open data cloud.

One RDF store in the cloud that caused a lot of attention after its launch is Neptune[10]. Due to a lack of descriptions that can be found, it is unclear which architecture it has.

3.1   RDF Stores Using Cloud Computing Frameworks

Implementing a distributed system is a challenging task. To reduce the complexity, several RDF stores in the cloud are realized on top of cloud computing frameworks. As shown in Fig. 4, these RDF stores need a master node that translates the RDF graph into some format that can be stored within the cloud computing frameworks. This is the job of the graph converter. Similarly, SPARQL queries need to be translated into queries or tasks that can be executed on the cloud computing framework to produce the query results that are then transferred back to the user. This is done by the query translator.

One of the first cloud computing frameworks that was used to build RDF stores in the cloud is Hadoop[11] [135]. RDF stores like SHARD [116], HadoopRDF [42] and CliqueSquare [44] transform the RDF graph into one or several files that are stored in the distributed file system of Hadoop [128]. SPARQL queries are translated into one or several jobs that are executed within Hadoop. Depending

---

[10] https://aws.amazon.com/neptune/.
[11] https://hadoop.apache.org/.

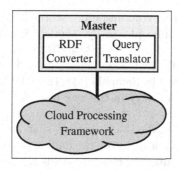

Fig. 4. Architecture of an RDF store using a cloud computing framework.

on how the RDF graph is separated into files, one or several files are processed during query execution.

To simplify the usage of Hadoop, the high-level query language and execution framework Pig[12] [99] was developed on top of Hadoop. Instead of translating SPARQL into Hadoop jobs directly, systems like PigSPARQL [122] and RAPID+ [72] translate SPARQL into the Pig query language. Pig then translate the code into Hadoop jobs and tries to optimized the orchestration of the individual jobs.

One of the main limitations of Hadoop is that the result of each individual task has to be written back into the distributed file system. To overcome this limitation, Spark (see footnote 1) [143] was developed. Spark stores the result of each job in main memory. These results can be used by several other jobs before the final result is optionally persisted on disk. Spark is used by SPARQLGX [50], S2RDF [124], SPARQL-Spark [96] and PRoST [29].

On top of Spark the graph processing framework GraphX[13] [45] was developed. In this framework a graph can be loaded and algorithms can be performed on it. These algorithms are vertex-centric, i.e., each vertex is able to receive, process and send messages to its neighboured vertices. S2X [121] translates SPARQL queries into such vertex-centric algorithms to produce the query results. Similar to S2X TripleRush [130], Random Walk TripleRush [132] (both basing on Signal/Collect [131]) and [46] use vertex-centric graph processing frameworks.

Another type of cloud computing frameworks are NoSQL (Not only SQL) database systems. These systems usually scale well with a high number of compute nodes and a fault tolerant. Their withdraw is that they usually do not provide ACID transactions. One type of NoSQL systems are key-value stores. These systems map a unique key to an arbitrary value. DynamoDB[14] [34] is such a distributed key-value store. It is used in AMADA [24] to index and store the triples of the RDF graph.

Column-family stores are another type of NoSQL database systems. These systems store tabular data like in relational databases. Instead of storing all data

---

[12] https://pig.apache.org/.

[13] https://spark.apache.org/graphx/.

[14] https://aws.amazon.com/de/dynamodb/.

of a row physically together, column-family stores locate all entries of a set of columns (i.e., a column-family) physically together. Examples of these stores are HBase (see footnote 2) (used by Jena-HBase [70], H_2RDF+ [103]), Cassandra[15] [77] (used by CumulusRDF [75]), Accumulo[16] (used by Rya [114] and RDF-4X [4]) and Impala[17] [117] (used by Sempala [111, 123]). RDF stores relying on these column-family stores usually vary in the way how they store the RDF graph in tables.

The last type of NoSQL database systems that are used in RDf stores are document stores. Document stores store the data in documents which are objects with arbitrary fields and values. Each of the fields can be used to index the data. The RDF store [32] uses Couchbase[18] and D-SPARQ [95] uses MongoDB[19].

3.2   Distributed RDF Stores

In contrast to RDF stores that use cloud computing frameworks, distributed RDF store have to address the challenges resulting from the distribution. To reduce the complexity of these challenges, most distributed RDF stores are realized with a master-slave architecture. In this architecture a dedicated compute node is the master. It is responsible for coordinating the individual slaves that store the RDF graph and process the queries. The disadvantage of this architecture is that the master node can easily become a bottleneck of the distributed RDF store. To overcome this limitation some distributed RDF stores are realized with a peer-to-peer architecture in which the design of every compute node is identical.

Distributed RDF Stores with Master-Slave Architecture. In distributed RDF stores that have a master-slave architecture there exists one master and several slaves. The master is a dedicate compute node that is responsible for the coordination of all slaves. The slaves are the compute nodes that store the graph and process the queries. The general architecture of such a distributed RDF store is shown in Fig. 5.

When a graph is loaded, the master first replaces each string identifier of a resource by a shorter unique integer identifier. This replacement reduces the storage size of the graph that needs to be processed. The mapping between the string and integer identifiers are stored in the dictionary. Thereafter, the graph cover creator assigns the triples to the individual slaves. Thereby, an index is created which keeps track on which slave which part of the graph is stored. Each slaves stores the triples assigned to him in its local RDF storage.

---

[15] https://cassandra.apache.org/.
[16] https://accumulo.apache.org/.
[17] https://impala.apache.org/.
[18] https://www.couchbase.com/.
[19] https://www.mongodb.com/.

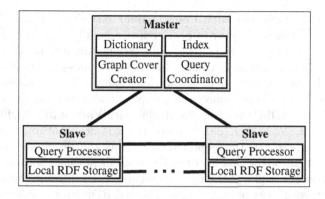

Fig. 5. Architecture of a master-slave distributed RDF store.

In order to query the RDF store, a query is send to the master. The query coordinator first encodes all string identifier in the query with the help of the dictionary. The query is then translated into a query execution tree and optimized. With the help of the index, the query coordinator can decide which part of the query can be executed on which slave and initiates the query processing on the slaves. On each slave the query processor processes the (sub)query assigned to him on the local RDF storage. The intermediate results can then be directly exchanged between all slaves. The final results are sent back to the query coordinator. With the help of the dictionary the query coordinator replaces the integer identifier of the results by it string identifier and sends them back to the user.

The architectures of Custered TDB [102], COSI [23], Partout [43], GraphDB [17], Blazegraph [2], SemStore [138], TriAD [52], DREAM [55,105], DiploCoud [104,139] are as described above. But there also exist variations of this architecture. For instance in Trinity.RDF [144], WARP [62], YARS2 [60] and 4store [58] the query coordinator also has to join intermediate results.

Some of the components of the master can be distributed among the slaves. For instance, in EAGRE [145] the graph cover creation is performed on all slaves in parallel and the index is distributed over all slaves. Furthermore, distributed RDF stores do not necessarily need all of the presented components. For instance, PHDStore [10], 2way [27,106] do not need a global index and/or dictionary. If distributed RDF stores adapt the graph cover during runtime based on the actual workload, the master also contains a redistribution controller as in AdPart [57], AdHash [9,56] and Spartex [5].

Beside these pure distributed RDF stores there also exist hybrid RDF stores that also use some cloud computing infrastructure. In Sedge [141] a reimplementation of Pregel [84] is used to process distributed queries. A more common approach is, to use Hadoop to process only distributed joins as done in [39,63,142], SPA [81], VB-Partitioner [79], SHAPE [80,137].

Distributed RDF Stores with Peer-to-peer Architecture. In distributed
RDF stores that follow the peer-to-peer architecture all compute nodes – called
peers – consist of the same components. This architecture has the advantage that
no single compute node can become a bottleneck by design. The disadvantage
is that usually no compute node knows how many compute nodes exist and
how the data is distributed among all compute nodes. Usually in systems like
Edutella [37,97], RDFPeers [25], PAGE [35], GridVine [6,31], RDFCube [87],
Atlas [67], 3RDF [12,13,101], a distributed index is used that routes requests to
the compute node storing the requested data. Since this distributed index is the
central component of the architecture, the implementation choice of the index
determines how the triples of the RDF graph are assigned to the compute nodes.
The architecture of most peer-to-peer distributed RDF stores is shown in Fig. 6.

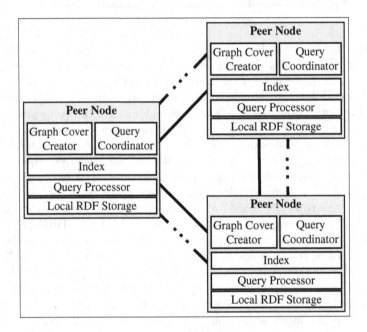

Fig. 6. Architecture of a peer-to-peer distributed RDF store.

To load an RDF graph, the complete graph can be sent to any compute
node. The graph cover creator asks the distributed index on which compute
node the individual triples should be created. Based on this decision the triples
are distributed over the compute nodes. When a compute node receives triples
from a graph cover creator, it inserts them into its local RDF store and updates
the distributed index. In order to speed up the graph loading procedure, the
RDF graph can be split into several parts that are processed by the graph cover
creators on several compute nodes in parallel.

When a user sends a query to any of the compute nodes, the query coordinator creates the query execution tree. Based on the index the query coordinator

decides which part of the query execution tree is sent to which compute node. When a query processor receives some subquery from a query coordinator, it retrieves the requested information from its local RDF store. The intermediate results are sent to other compute nodes or back to the query coordinator where they will be further processed. Finally, the query coordinator computes the overall results and send them back to the user.

In contrast to the master-slave architecture, peer-to-peer distributed RDF stores usually do not have a global dictionary. As a consequence string identifiers are used when transferring triples during the loading of a graph or intermediate results during the query processing.

3.3   Federated RDF Stores

In the linked open data cloud[20] there exist many public datasets and SPARQL endpoints. In order to combine several datasets and query them together a naive approach would be to download all datasets and load them into a single RDF store. This naive approach has several disadvantages. First of all, it requires a lot of computational resources to store and process several datasets at once. Furthermore, when the datasets change, the system would need to keep track of these changes and update its local copies of the datasets.

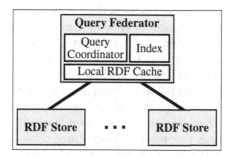

Fig. 7. Architecture of a federated RDF store.

To overcome these limitations, federated RDF stores have been developed. Their basic idea is to query remote RDF stores directly. The general architecture of federated RDF stores is shown in Fig. 7. In order to query the remote RDF stores a global index is required that indicates which RDF stores contain data that are relevant for the query. This index is created by retrieving statistical information that are provided by the remote RDF stores (e.g., SPLENDID [48], WoDQA [8], LHD [134], DAW [120], SemaGrow [26], FEDRA [91], LILAC [92] and Odyssey [90]) or the user (e.g., DARQ [115]), by sending special queries to the remote RDF stores (e.g., FedX [127], ANAPSID [7,93], Lusail [86]) or by

---

[20] http://lod-cloud.net/.

observing the results that are returned during the processing of user queries (e.g., ADERIS [83]). Also combinations of these strategies are possible as in Avalanche [18]. Also the absence of the index is possible if the resource IRIs are dereferenced as proposed by SIHJoin [76].

When a user sends a query, the query coordinator transforms it into a query execution tree. With the help of the index it can decide from which remote RDF store can contributed to the query. Based on this decision it forwards the query or parts of it to the corresponding remote RDF stores. The returned intermediate results are joined by the query coordinator and finally sent back to the user. In order to speed up the query execution, the query federator also contains a local RDF cache in which data retrieved by previous queries is cached. This cached data can be reused for future queries, which might reduce the number of queries that needs to be sent to the remote RDF stores.

4   Graph Cover Strategies

One core aspect of RDF stores in the cloud is, how the RDF graph is distributed among the compute nodes, resulting in a graph cover. A common procedure to create a graph cover is to first split the RDF graph into small possibly overlapping subsets. Thereafter, this graph subsets are assigned to compute nodes. In RDF stores that bases on cloud computing frameworks, the influence on how this assignment to compute nodes is done is usually limited. Therefore, in Sect. 4.1 is described how a graph is split into subsets that are then stored in the cloud computing framework. The graph cover strategies of distributed RDF stores can in general be separated into three categories: (i) hash-based graph cover strategies (see Sect. 4.2), (ii) graph-clustering-based graph cover strategies (see Sect. 4.3) and (iii) workload-aware graph cover strategies that distributes the graph based on a historic query workload (see Sect. 4.4). In order to reduce the amount of queries that need to combine data from different compute nodes, the n-hop replication was proposed that replicates triples at the border of the chunks of arbitrary graph cover strategies (see Sect. 4.5).

When RDF stores are queried frequently, the initial distribution of the graph on the compute nodes might be suboptimal for the current query workload. To improve the query performance, some RDF stores have implemented a dynamic graph cover strategy (see Sect. 4.6). This strategy observes the current query workload and tries to optimize the data placement by moving or copying parts small triple sets from one compute node to another.

Since there are is huge number of graph cover strategies, we focus in this section only on the most frequently graph cover strategies and give only hints to a few variations that can be found. We do not present exotic graph cover strategies that were only used in a single RDF store. Parts of this section was taken from [64].

4.1 Graph Cover Strategies in Cloud-Computing-Framework-Based RDF Stores

RDF stores that build on cloud computing frameworks have usually limited influence how the data is placed on the individual RDF stores. There influence is limited to the way how the RDF graph is split into subsets that are stored in files or tables. The goal of splitting the RDF graph into subsets is to reduce the amount of triples that need to be processed during the query execution. This is achieved by storing all triples with the same subject, object, predicate or combinations of them in the same file or table.

Molecule Graph Splits. In order to process star-shaped queries efficiently, the RDF graph is split into molecules of diameter 1. This means that all triples with the same subject are stored in one file (Fig. 8). D-SPARQ [95], follows this approach. The advantage of the molecule graph cover is that star shaped queries whose triple patterns are joined on a subject no join needs to be processed. In case of a constant subject only a single file needs to be processed.

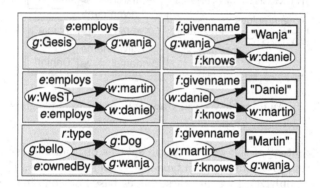

Fig. 8. The example graph split into molecules.

In order to reduce the number of required joins, RAPID+ [72] proposes to store all triples that have the same resource identifier at a subject or object position in a single file. Additionally, RAPID+ reduces the number of joins by increasing the molecule diameter in order to process path-shaped queries more efficiently.

In some RDF stores like Stratustore [129], Sempala [123] and SHARD [116] all molecules are stored in a single table. Each molecule is basically represented by a single row in this table with the subject as unique identifier. The predicates are the column names and the object as the value stored in each cell. If the combination of subject and property occurs in several triples, these cells store a list of objects or several rows for this subject are created. This storage layout is called property table in the literature.

Vertical Graph Splits. The basic idea of the vertical cover originated in [3] to store RDF data in a relational database so that for each property a table is created in which all triples with this property are stored. In the context of distributed RDF stores, approaches like Jena-HBase [71], PigSPARQL [122,146] and SPARQLGX [124] store all triples with the same property in a file or table (Fig. 9). The advantage is that it is easy to compute but a query that matches with paths of length l will only match with triples on at most l compute nodes. Thus, this graph cover strategy is likely to result in an imbalanced workload and a high number of exchanged intermediate results.

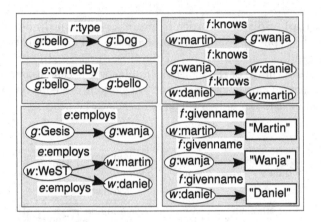

Fig. 9. An example vertical graph split of the example graph.

One disadvantage of the vertical graph split as presented above is, that frequently occurring properties like rdf:type lead to very large files. Therefore, the RDF store HadoopRDF [42] splits these tables based on combinations of properties and the RDFS types of the objects.

Another variant of the vertical graph split is realized in S2RDF [124]. In order to reduce the number of joins that need to be processed, they additionally create tables for all possible subject-subject and subject-object joins of triples.

4.2 Hash-Based Graph Cover Strategies

Hash Cover. A hash cover assigns triples to chunks according to the hash value computed on their subjects modulo the number of compute nodes. Thus, all triples with the same subject – i.e., a molecule – are located in the same graph chunk. This graph cover strategy is used, for instance, by Virtuoso Clustered Edition [40], VB-Partitioner [79], SPA [81], 4store [58] and AdHash [9].

Example 9. The following hash function produces the graph cover shown in Fig. 2.

$$\forall r \in \{\text{g:Gesis, g:wanja, w:martin}\}: \text{hash}(r):= 1$$
$$\forall r \in \{\text{g:bello, w:WeST, w:daniel}\}: \text{hash}(r):= 2 \ .$$

The advantages of the hash cover are that it is easy to compute and due to a relatively random assignment of triples to compute nodes the resulting graph chunks will have similar sizes. The disadvantages are that it may lead to a high number of exchanged intermediate results if a query matches with long paths. Since all hash covers are subject-contained, this graph cover strategy might be a good choice if the expected queries will only match with paths of a short length (ideally 1).

Beside the subject, distributed RDF stores like Trinity.RDF [144], Clustered TDB [102], YARS2 [59,60] and RDFPeers [25] also use property and/or the object to assign each triple three times to the compute nodes.[21] Additionally, RDF stores like PAGE [13,35] append at least two elements of each triple and use the hash of the result to assign triples to compute nodes.

Hierarchical Hash Cover. Inspired by the observations that IRIs have a path hierarchy and IRIs with a common hierarchy prefix are often queried together, SHAPE [80] uses an improved hashing strategy to reduce the inter-chunk queries. First, it extracts the path hierarchies of all IRIs. For instance, the extracted path hierarchy of "http://www.w3.org/1999/02/22-rdf-syntax-ns#type" is "`org/w3/www/1999/02/22-rdf-syntax-ns/type`". Then, for each level in the path hierarchy (e. g., "org", "org/w3", "org/w3/www", ...) it computes the percentage of triples sharing a hierarchy prefix. If the percentage exceeds an empirically defined threshold and the number of prefixes is equal or greater to the number of compute nodes at any hierarchy level, then these prefixes are used for the hash cover.

Example 10. Assume the hash is computed on the prefixes gesis and west of the subject IRIs in the example graph. If the hash function returns 1 for gesis and 2 for west the resulting hierarchical hash cover is shown in Fig. 10.

In comparison to the hash cover the creation of a hierarchical hash cover requires a higher computational effort to determine the IRI prefixes on which the hash is computed. For queries that match with paths in which the subjects and objects have the same IRI prefix the number of exchanged intermediate results may be reduced. This reduction might come at the cost of a more imbalanced query workload since only a few chunks will contain these paths. Thus, the use of the hierarchical hash cover might be beneficial (i) if the network connecting the compute nodes is slow or (ii) if other functionality such as prefix matching benefits from the hierarchical hash cover.

---

[21] If the hash cover is only computed on the predicate, the resulting graph cover would be similar to the vertical graph split.

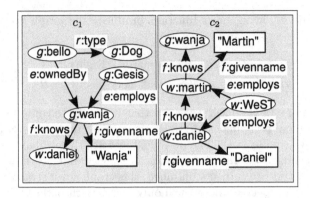

Fig. 10. An example hierarchical hash cover which is also a minimal edge-cut cover of the example graph.

4.3   Graph-Clustering-Based Graph Cover Strategies

Graph clustering considers splitting a graph into partitions, i.e., a graph cover with pairwise disjoint graph chunks. In this area a wide variety of algorithms were developed (for instance, see the survey [85]). The basic idea is that an RDF graph is partitioned by one of these algorithms. Since most of the graph clustering algorithm create an assignment from vertices to compute nodes, the triple are usually assigned to the compute node to which its subject was assigned to. The most frequently applied graph clustering approach is the minimal edge-cut partitioning which described below.

Another rarely used graph clustering algorithm tries to optimize the modularity. The modularity measures the difference between the actual number of edges within the partitions and the expected number of such edges. This strategy was applied for instance by MO+ [113].

Minimal Edge-Cut Cover. The minimal edge-cut cover is a vertex-centred partitioning which tries to solve the k-way graph partitioning problem as described in [69]. It aims at minimizing the number of edges between vertices of different partitions under the condition that each partition contains approximately $\frac{|V_G|}{k}$ many vertices. Details about the computation of k-way graph partitioning and the targeted approximation can, e.g., be found in [69]. RDF stores like D-SPARQ [63,95,105] convert the outcome of the minimal edge-cut algorithm, i.e., a partitioning of V_G, into a graph cover of G by assigning each triple to the compute node to which its subject has been assigned.

Example 11. A minimal edge-cut algorithm might assign the resources g:Dog, g:Gesis, g:bello, g:wanja and "Wanja" to compute node c_1 and all other resources to compute node c_2. For our specific running example the result of the minimal edge-cut cover strategy is identical to the results of the hierarchical hash cover strategy depicted in Fig. 10.

In this example there exist two edges connecting vertices assigned to different chunks. One is the f:knows edge starting at g:wanja and ending at w:daniel. The other is the f:knows edge starting at w:martin and ending at g:wanja. Since the subject g:wanja of the first triple is assigned to c_1, this triple is assigned to c_1. The subject of the second triple w:martin is assigned to c_2. Therefore, this triple is assigned to c_2.

Since the minimal edge-cut cover considers the graph structure, the creation of the graph cover requires a high computational effort. The advantage of considering the graph structure might be a reduced number of exchanged intermediate results. This would make the minimal edge-cut cover a good choice if the network connection between compute nodes is slow.

In order to optimize the query performance, TriAD [52] creates an over-partitioning. For instance, to create a graph cover that assigns triples to 5 compute nodes, 100k-200k partitions are created. These partitions are then assigned to compute nodes. To improve the performance of queries that use RDFS schema information, in [108] the RDFS schema is replicated to all graph chunks. Since the minimal edge-cut can lead to graph chunks whose cardinally vary strongly, [133] proposes to weight vertices by the number of triples in which it occurs.

An alternative optimization is performed by EAGRE [145]. Instead of partitioning the original graph, a minimal edge-cut cover of the summary graph is created. Each vertex of the summary graph represents a set of molecules that have similar predicates. An edge (v_1, v_2) in the summary graph is created, if any anchor vertex of the molecules contained in v_2 occur as object in any molecule contained in v_1. The vertices are weighted by the number of molecules they contain. This graph cover strategy ensures that molecules with similar predicates are stored on the same compute node.

4.4   Workload-Aware Graph Cover Strategies

Another type of graph cover strategies assume that the query workload does not change much over the time. Therefore, they learn from a historic query workload which triples have been frequently queried together first. Based on this knowledge they try to find a optimal graph cover for future queries. These approaches are, for instance:

- The novel idea applied in WARP [62] is creating an initial minimal edge-cut cover and then replicate triples in a way such that all historic queries can be answered locally.
- In COSI [23] edges are weighted based on the frequency they are requested by the historic query workload. Thereafter, a weighted minimal edge-cut partitioning is performed leading to an improved horizontal containment.
- In [17] the resulting graph cover aims to balance the overall workload of all queries equally among all compute nodes. Thereby, each query is processed by a single compute node in an ideal case. To reach this goal, the proposed algorithm assigns the triples required by the queries to compute nodes in a way that the number of replicated triples is reduced.

- In Partout [43] the queries contained in the historic query workload are first generalized by replacing every rarely queried subject or object constants by variables. Thereafter, the matches of this generalized triple patterns are assigned to compute nodes in a way that (i) ideally each query can be answered by a single compute node without replicating triples and (ii) the query workload of all queries is distributed equally among all compute nodes.
- DiploCloud [139] generalizes the queries in the historic query workload by using schema information. Then triple sets are computed that can produce a single query result. Finally, the triple sets are distributed equally among all compute nodes.

4.5 n-Hop Replication

Whenever a query combines data from different graph chunks, intermediate results need to be exchanged between different compute nodes. To reduce the number of exchanged intermediate results for a subject-complete graph cover of graph G, the n-hop replication strategy extends each of its chunks ch_i by replicating all triples contained in some path of length $\leq n$ in G starting at some subject or object occurring in ch_i. This way all queries that match with paths of length $\leq n$ could be processed without exchanging intermediate results. The n-hop replication is used by systems like [63], VB-Partitioner [79] and D-SPARQ [95].

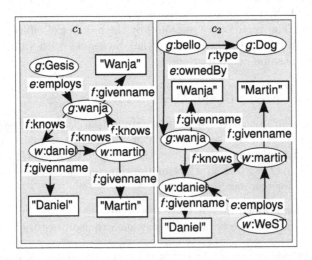

Fig. 11. The 2-hop extension of the hash cover in Fig. 2.

Example 12. Applying the 2-hop replication extension on the hash cover in Fig. 2 results in the 2-hop hash cover shown in Fig. 11. In this cover a query could match with the path (g:bello, e:ownedBy, g:wanja) , (g:wanja, f:knows, w:daniel) on compute node c_2 without the need to exchange intermediate results.

The n-hop replication may reduce the number of transferred intermediate results at the cost of replicating triples. This replication will increase the effort to create the graph cover and increase the size of the graph chunks. Furthermore, the replication might cause a higher computational effort during the query processing since the replicated triples might lead to duplicate intermediate results. Thus, using the n-hop replication might be beneficial if the network connecting the intermediate results is slow and the number of replicated triples is low.

4.6   Dynamic Graph Cover Strategies

Graph covers created by one of the graph cover strategies above can lead to a high amount of data transfer between compute nodes, if the actual query workload needs to combine data stored on different compute nodes. In order to overcome this limitation, PHD-Store [9] and AdHash [9] keeps track of basic graph patterns that are queried frequently. When the frequency exceeds a threshold, triples that match with these frequent triple patterns are replicated in a way that these basic graph patterns can be executed locally.

Instead of only trying to reduce the network communication, Sedge [9] tries to primarily distribute the query workload equally among the compute nodes. Therefore, Sedge keeps track how frequently the molecules are queried together. If a set of molecules are queried together with a high frequency, these set of molecules is replicated to a compute node with a low workload.

Another type of graph cover strategies assumes that during runtime new triples can be added to the RDF store. In this setting it may happen that a single compute node stores much more triples then other compute nodes. To prevent the compute node from being overloaded, the triples of that compute node can be redistributed based on the prefix of some hash values (as done in [101]). Another strategy is performed by [19]. The triples are sorted lexicographically and one half of them is sent to another compute node. In both cases the systems keep track to which compute node they have moved the triples.

5   Indices

RDF stores in the cloud distributed the triples of an RDF graph over several computers. When a query is sent to these RDF stores, they require an index which can tell on which compute nodes the data contributing to the query processing is located. These indices are either stored on a single compute node – i.e., the master or the query federator – (see Sect. 5.1) or they are distributed over several compute nodes (see Sect. 5.2). A centralized index has the advantage that it has knowledge about all graph chunks. To reduce the size of the index, some type of aggregation needs to be applied. In contrast to this distributed indices need fewer aggregation, since they are stored across all compute nodes. But a single index lookup might require the routing of the lookup via several compute nodes until the required information is found.

5.1   Centralized Indices

Hash-Based Index. In distributed RDF stores that apply some variant of a hash cover strategy (for instance, Virtuoso Clustered Edition [40], 4store [58] or Trinity.RDF [144]), no explicit index is required. Based on the knowledge of all compute nodes, the hash function and the triple elements that were hashed, the compute node to which triples following a specific pattern were assigned can be identified.

Statistics-Based Index. Another type of centralized indices base on statistical information about the resources occurring in the individual graph chunks. In RDF stores like DARQ [115], FedX, [133] and Sedge [141] the frequency of every subject, property and object in each chunk is counted and stored. Since these information do not tell anything about the RDFS types contained in the graph chunks, systems like SPLENDID [115], WoDQA [115], LHD [115], SemaGrow [115], Avalanche [115] bases on VoID descriptions [11] of each graph chunk. These descriptions contain the occurences of URIs in the dataset, the used RDFs types and the properties that occur in triples whose objects occur as subject in triples assigned to a different compute node. Since these information might be complicated to collect in a federated setting, if the remote RDF stores do not provide VoID descriptions, ANAPSID [115] restricts itself to only count the occurrences of properties and RDFS types.

If a triple pattern with two constants is requested, the indices described so far could only restrict the number of queried compute nodes by either of the two constants. To restrict the number of queried compute nodes even further, LILAC [92], SemStore [138] additionally count how frequently all subject-property, property-object and subject-object combinations occur.

Since not all subjects and objects occurring in a dataset have an RDFS type, the RDF store Odyssey [90] defines the type of an subject v by the set of properties occurring in its molecule M_v. Since the number of types might me very large, types with similar property sets are merged. Additionally, it counts how frequently instances of molecule types are connected by properties.

Summary-Graph-Based Index. In distributed RDF stores summary graphs are created. This summary graph can be queried to identify compute nodes that store triples required for the processing of a query. The definition of a summary graph differ between the RDF stores.

In TriAD [52] each graph chunk becomes a vertex in the summary graph. Since TriAD uses minimal edge-cut cover, the underlying algorithm assigns each vertex to a compute node. For each triple (s, p, o), an edge with label p is crated in the summary graph that connects the vertices representing the chunks stored on the compute nodes to which s and o were assigned to. To reduce the size of the summary graph, all edges connecting the same vertices and having the same label are represented by only a single edge.

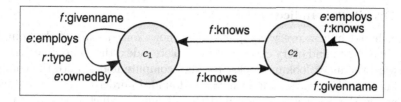

Fig. 12. The summary graph of the graph cover shown in Fig. 10 used by TriAD.

Example 13. Figure 12 shows the summary graph created from the minimal edge cut cover in Fig. 10. The vertices represent the compute nodes. For instance, c_1 represents the graph chunk of compute node c_1. The self-loop at the vertices represent the properties occurring within these compute nodes. To simplify the graphic, all labels were attached to a single edge instead of creating an own edge for every label. The two edges connecting both vertices represent both triples whose subject and object were assigned to different compute nodes.

Another type of summary graph is used by EAGRE [145]. Vertices in the summery graph represent molecule types. A molecule type is a set of all properties that occur in at least one molecule. To reduce the number of molecule types, molecule types with similar properties are merged. Each triple t contained in a molecule of type T_1 whose object is the anchor vertex of another molecule of type T_2 will result in an edge with the label of the property that connects the vertices T_1 and T_2 of the summary graph.

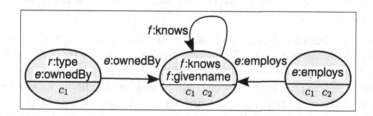

Fig. 13. The summary graph of the graph cover shown in Fig. 10 used by EAGRE.

Example 14. Figure 13 shows the summary graph created by EAGRE for the graph cover shown in Fig. 10. The three vertices represent the three different molecule types. The properties occurring in molecules of that type are written in the upper part of the vertex. The compute nodes on which molecules of that type can be found are listed in the bottom part of each vertex. The leftmost vertex represents the dog molecule, the middle vertex represents the employee molecules and the right vertex represents the institutes molecules. The e:employs edge connecting the institutes molecule type with the employees molecule type represents the edges that connect the two institutes with their employees.

5.2 Distributed Indices

In distributed indices every compute node knows only a part of the complete index. In order to find every entry of the complete index, the compute nodes have to forward the index lookup request to other compute nodes, until the compute node knowing the requested information is found. In order to route these index lookups, overlay networks are created that define to which compute node an index lookup should be forwarded.

Hash-Based Index. If a distributed RDF store uses a hash partitioning, each compute node can determine on which compute node a triple can be found, by knowing the set of all available compute nodes. This approach is done, for instance, by HDRS [22] and Virtuoso Clustered Edition [41].

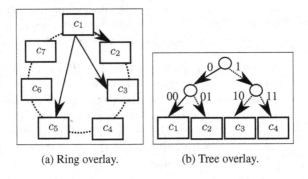

(a) Ring overlay.   (b) Tree overlay.

Fig. 14. The different types of overlay networks used in distributed hash indices.

In peer-to-peer distributed RDF stores, the set of all compute nodes might be large and change over time. To prevent the replication of this set and keeping it up-to-date on all compute nodes, each compute node only stores a set of neighboured compute nodes, to which index lookups might be forwarded. The definition of the neighbourhood creates an overlay structure. In peer-to-peer distributed RDF stores the following overlay structures are used frequently:

Ring structure. In RDFPeers [25], PAGE [35], Atlas [67] and [82] the compute nodes are order, e.g., by their IP address. This order defines the direct neighbours of each compute node. To ensure that every compute node has exactly two neighbours, the first and last compute node are defined as neighbours. An example of the resulting ring is shown in Fig. 14a. If only the ring structure would be given, finding a compute node that stores the requested data would take linear time. To reduce the lookup time, each compute nodes stores shortcuts to compute nodes at later positions in the ring. For instance, compute node c_1 knows compute nodes c_2, c_3 and c_5. If c_1 is asked whether it knows some information about resource r, it can compute with the hash that triples with this resource would have been assigned to, e.g., c_6. Since he does not know c_6 he sends the request to c_5 which is closest to c_6.

Tree structure. In GridVine [6,31], UniStore [68], 3RDF [12–14] the overlay network is based on a prefix tree as shown in Fig. 14b. Each vertex in this tree represents a prefix. The root has an empty prefix, the left child of the root node has the prefix 0 and the leaf c_1 stores all triples with resources whose hash value start with 00. Each compute node knows the path from the root to itself. For each node n in the path, the compute node knows one compute node contained in the subtree of the siblings of n. For instance c_1, would forward every hash with prefix 01 to compute node c_2 and every hash with prefix 1 to compute node c_3.

Combinations of both overlay structures can be found in [19,101]. The basic idea is initially they use the ring overlay structure. If one compute nodes has to store too many entries, it redistributes it triples based on a tree structure.

Schema-Based Index. Instead of using hash-based indices, the RDFS types can be used to distribute data. The RDFS types contained in a dataset usually build a type hierarchy. Similar to the tree overlay structure presented above, each compute node is responsible for the instances of the RDFS types assigned to him. If a compute node should retrieve instances of a given type T that is not assigned to him, it searches for a superclass of T for which he knows a responsible compute node and forwards the request to it. This so called semantic overlay network [30] is used, for instance, by SQPeer [73].

Chunk-Integrated Summary Graph Index. A completely different type of distributed index is presented in [108,109]. It adapts the idea of the summary graph from TriAD as described in Sect. 5.1. Instead of realizing an separate index structure, it integrates the summary graph into its local RDF storage. With the help of these additional information a compute node can decide, whether the triples of another compute node has data that might lead to further query results.

Example 15. Figure 15 shows the minimal edge-cut graph cover from Fig. 10 extended by the triples from the summary graph. On c_1 the graph chunk from c_2 is represented by a super vertex named c_2. All triples whose subject or object is contained in the graph chunk of c_1 but the counterpart not, is represented by an edge connecting the vertex within the chunk of c_1 with the super vertex representing the graph chunk in which the other vertex is contained. For instance, the subject of the triple `w:daniel f:knows w:martin` is contained in the chunk of c_1 whereas its object is only contained in the chunk of c_2. Therefore, the chunk of c_1 is extended by the triple `w:daniel f:knows` c_2. Furthermore, for each property p label occurring in the graph chunk of c_2, a triple c_2 p c_2 is added to the chunk of compute node c_1.

6   Distributed Query Processing Strategies

RDF stores in the cloud distribute RDF graphs over several compute nodes. One challenge which arises from this distribution is how to query the distributed

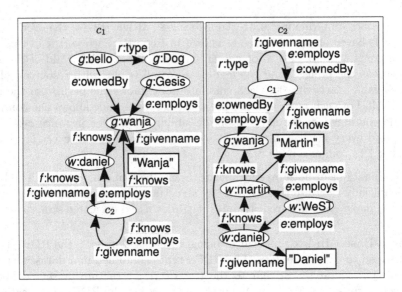

Fig. 15. The summary graph integrated into the graph cover shown in Fig. 10.

graph. In general RDF stores in the cloud try to compute as much on the graph chunks as possible without the need to exchange data. Therefore, the received query is decomposed into subqueries that can be executed only on the individual graph chunks stored on a single compute node. In the simplest case these subqueries consist of a single triple pattern. Other systems can make use of some properties of the underlying graph cover. For instance, SHARD [103] uses a graph cover that assigns all triples with the same subject to one compute node. As a result, all star-shaped subqueries can be executed locally. Another example are RDF stores that make use of the n-hop replication. This replication ensures that all queries that match with a subgraph with diameter n can be executed locally. If the local RDF storage is able to return all partial matches of the query, the complete query can be executed by the local RDF storage. With the help of the indices, the number of compute nodes that can contributed to a subquery can be restricted.

The intermediate results of the subqueries needs to be joined in order to produce the overall query result. Several RDF stores transfer all intermediate results to a single compute node that is then responsible for joining it (see Sect. 6.1). If a huge number of intermediate results are produced, the joining compute node might be overloaded. Therefore, several RDF stores distribute the join computation over several compute nodes (see Sect. 6.2). An example how queries are executed in distributed graph processing frameworks is given in Sect. 6.3.

6.1   Centralized Join

Especially in federated RDF stores, the intermediate results of subqueries that were executed on remote RDF stores have to be joined on the query federator. Thereby, the join strategies of relational databases are applied. The following join strategies are used:

Nested loop join [89]: For each element in the intermediate result list of the first subquery, the intermediate result list of the second result list needs to be iterated completely to find all join candidates.

Merge join [89]: For this join the intermediate result lists must be ordered. One list is iterated and for each element the join candidates in the other list can be retrieved. Due to the ordering the other list does not need to be traversed from the beginning. Instead only the elements with the same value of the join variable needs to be reiterated. This join is performed by the distributed RDF store Partout [43].

Hash join [36]: The intermediate results of the subqueries are stored in separate hash tables. Each hash table distributes the results into several buckets. If both hash tables use the same number of buckets, only the intermediate results of two buckets need to be joined at once. When all buckets are joined, the join is finished.

Symmetric join [136]: This type of join is a non-blocking hash join. For every subquery a hash table is created that stores the already received intermediate results. When a new intermediate result is received, it is joined with all join candidates in the other hash tables. The results are emitted and the intermediate result is inserted in the hash table of the subquery that produced it. This join strategy is performed by, e.g., ANAPSID [7] and LHD [134].

Bind join [54]: In order to perform a bind join, the first subquery is executed. For each returned distinct intermediate variable binding μ, the second subquery is executed. Thereby, all variables bound by μ in the second subquery are substituted by the bound values. This type of join is performed by, e.g., FedX [127], Avalanche [18] and SemaGrow [26].

Beside performing a single type of join operation, RDF stores like DARQ [115] and SPLENDID [48] choose between a bind join and a nested loop join or between a bind join and a hash join, respectively. The choice depends on the expected number of returned results.

Another join strategy is performed by [105]. In this distributed RDF store partial evaluation [65] is performed. This means that the complete query is executed on the local RDF stores of each compute node. These local RDF stores return the overall results and all intermediate results. For each intermediate result the subquery that created the result is returned. The intermediate results from all compute nodes are sent to a single compute node who finally joins the intermediate results and returns the overall results.

6.2   Decentralized Join

The subqueries in which a query is decomposed can easily produce a large number of intermediate results. Joining all intermediate results on a single compute node can overload the capacity of this compute node. To overcome this limitation several RDF stores in the cloud apply distributed joins.

Replication-Based Distributed Join. In order to distribute the number of join computations over the individual compute nodes, in SemStore [138] all intermediate results of a subquery are sent to all compute nodes on which the succeeding subquery is executed. This strategy increase the amount of transferred intermediate results but each compute node only joins its local results with the intermediate results produced by the other compute nodes.

Distributed Hash Join. In DiploCoud [139] the intermediate results of the subqueries are joined by a distributed hash join. Basically, the distributed hash join is similar to a centralized hash join. The only difference is that each compute node is a bucket of the used hash table.

The hypercube hash join was initially presented in [20] and was used in the distributed RDF store presented in [28]. The basic idea is that for each join variable one dimension is created. For instance, if a query has three join variable, three dimensions are created. Therefore, we need to arrange the compute nodes as a three-dimensional cube. Each compute node is responsible for one cubic region within in this cube. Thereafter, every triple pattern is executed in parallel producing variable bindings. If the variable binding $\{(?v1, w:martin)\}$ is produced, it is forwarded to all compute nodes that are responsible for the value hash($w:martin$) in the $?v1$-dimension. Each compute nodes performs a local join of the variable bindings it has received from the different triple patterns. Depending on the selection of the regions each compute node is responsible, the workload can be equally distributed among all compute nodes.

Distributed Merge Join. In a distributed merge join the intermediate result lists of the subqueries are sorted by the values of the join variables. Thereafter, each compute node receives all elements within a specific value range and joins them. This type of join is primarily performed in Hadoop-based RDF stores like H2RDF+ [103], SHARD [63,103] and Spark-based RDF stores like SparkRDF [140] and SPARQLGX [50].

Distributed Bind Join. On way to realize a distributed bind join is implemented in AdHash [9]. The first triple pattern is executed on each compute node. Thereafter, each compute node performs a centralized bind join based on the variable binding the first triple pattern has produced on this compute node.

In RDFPeers [25], GridVine [6,31], Atlas [67] and 3RDF [12] a hash cover is applied which each triple is stored on at most three compute nodes based on the hash of its subject, property and object. As a consequence all triples in which one resource occurs are located on the same compute node. Since usually each triple pattern of a query contains at least one constant, all matches for this triple pattern can be found on a single compute node. In order to process a query, the query coordinator determines a sequence in which the compute nodes should be traversed to process all triple patterns. When moving from one compute node to the other, all intermediate results are transferred. In order to join the intermediate results, bind joins are performed.

In order to generalize and parallelize the previously described strategy, [82] introduced the so called spread by value querying strategy[22]. The basic idea is that the first triple pattern is processed on the computed nodes on which matches occur. These compute nodes start a bind join processing with the second triple pattern. When a compute node identifies with the help of the global index that for one triple pattern there exist matches on a different compute node, it will fork the query processing on the other compute node that will continue with this branch of the query execution. The final query results are sent back to the query coordinator. This strategy was also used by, for instance, [13,14,108,109] and TripleRush [130]. In order to speed up the query execution, Trinity.RDF [144] performs the spread by value strategy from the first and the last triple pattern in parallel.

Example 16. In this example the basic graph pattern `<w:WeST> <e:employs>` `?v1. ?v1 <f:knows> ?v2. ?v2 <f:givenname> ?v3` should be executed on the graph cover with integrated summary graph index in Fig. 15. The first triple pattern creates the variable binding $\{(?v1, w:martin)\}$ on compute node c_2. Based on this variable binding the variable ?v1 of the second triple pattern will be substituted by `w:martin`. When processing the triple pattern `<w:martin> <f:knows> ?v2`, the variable binding $\{(?v1, w:martin), (?v2, g:wanja)\}$ is produced. Before processing the third triple pattern, ?v3 will be substituted by `g:wanja`. As a result `<g:wanja> <f:givenname> ?v3` is executed. This time the only possible substitution for ?v3 is the super vertex c_1. This match means that there exists a triple on compute node c_1 that would match with the triple pattern. Therefore, the query, the created variable binding, and the metadata that this variable binding was created by the first two triple pattern is sent to compute node c_1. Now, c_1 executes `<g:wanja> <f:givenname> ?v3` and produces the resulting variable binding $\{(?v1, w:martin), (?v2, g:wanja), (?v3, "Wanja")\}$. For the sake of simplicity, the other intermediate results produced by the triple patterns that were executed in parallel were skipped during the explanation of this example.

6.3   Distributed Query Processing in Graph Processing Frameworks

In S2X [140] a different type of distributed query processing is presented. First, each vertex checks whether it can be a substitution for some variable in the query

---

[22] This idea is named differently in the literature. For instance, in Trinity.RDF [144] it is called graph exploration.

by checking its incident edges, their labels and the adjacent vertices. Thereafter, it notifies the neighboured vertices by its variable bindings. If for one variable binding no join compatible variable binding can be found on the neighboured vertices, it is discarded. The notification of the neighboured vertices and the discard of local variable bindings is repeated until the variable binding of each vertex in the graph does not change any more. The remaining variable binding can be retrieved and joined as the final results.

7 Fault Tolerance

One problem of RDF stores in the cloud is that a single compute node might fail or become disconnected from the network. RDF stores that bases on cloud computing frameworks mainly rely on the fault tolerance of the used framework. In federated RDF stores the actual data is stored on remote RDF stores that are not under control of the system administrator. As a result the fault tolerance is not an urgent problem for both types of RDF stores.

In contrast to these RDF stores, the failure of compute nodes is an issue for distributed RDF stores. Since most of these RDF stores that can be found in the literature are proof-of-concept implementations, they do not address the problem of fault tolerance. The few systems that deal with fault tolerance, address this problem by replication. Systems like Virtuoso Clustered Edition [41] suggest to create identical copies of all compute nodes. If one compute node fails, it is replaced by one of its copies.

Another strategy to become fault tolerant is used by, e.g., 4store [58] and RDFPeers [25]. In these distributed RDF stores there exists an partial order of all compute nodes, for instance, created by the comparison of their IP addresses. To ensures that every compute node has a successor and predecessor, the successor of the last compute node is the first compute node. Based on this ordering, the triples assigned to one compute node are also assigned to the neighboured compute nodes. If one compute node fails, the index forwards the queries to one of neighbours that store replicas of the data originally assigned to the failed compute node.

8 Evaluation Methodologies

In order to evaluate the performance of RDF stores several benchmarks were proposed. In general benchmarks consist of a dataset, a set of queries and several performance metrics. In order to test the RDF stores with differently-sized RDF graphs, benchmarks usually use a dataset generator that generated RDF stores based on a schema and/or specific characteristics. Some benchmarks provide fixed queries or query patterns that contain special variables that are substituted by constants after dataset generation (see Sect. 8.1). Instead of providing query patterns, benchmark generators generate queries based on query characteristics (see Sect. 8.2). In Sect. 8.3, we elaborate how benchmark are used to evaluate distributed RDF stores.

8.1   Benchmarks

Lehigh University Benchmark. LUBM [51] was developed to test the query optimizer performance. It generates an RDF graph based on its Univ-Bench ontology. This ontology describes universities, their departments, employees, courses, students and related activities. In order to provide more realistic datasets, several constraints are applied during data generation. For instance, a university can have between 15 and 25 departments and the ratio between undergraduate students and faculty is between 8 and 14. The 14 provided SPARQL queries are designed to test how well the query optimizer can improve the join ordering. The characteristics of the queries are shown in Table 1. LUBM proposes the following performance metrics:

Table 1. LUBM query characteristics.

Query	# Triple patterns	Query diameter
Q1	2	1
Q2	6	2
Q3	2	1
Q4	5	1
Q5	2	1
Q6	1	1
Q7	4	2
Q8	5	2
Q9	6	2
Q10	2	1
Q11	2	1
Q12	4	2
Q13	2	2
Q14	1	1

Load time is the time that the RDF store needs to parse and load the RDF graph.

Repository size is the size of all files that are required by the RDF store to store the dataset including dictionary and indices.

Query execution time is the average time to execute a query ten times.

Query completeness and soundness measures with percentage of all results were retrieved and the percentage of correct results.

SP²Bench. SP²Bench [126] was designed to test the most common SPARQL constructs and how they are applied in realistic queries. It provides a dataset generator that creates an RDF graph that follows the DBLP schema. This schema describes publications like articles and inproceedings with their bibliographic information. The generated graph mimics the characteristics of the real DBLP graph. SP$_2$Bench provides 17 queries. They mainly focus on testing the join ordering capabilities of the query optimizer but also complex filters and duplicate elimination. The characteristics of the queries are given in Table 2. The proposed performance metrics are:

Table 2. SP^{2}Bench query characteristics.

Query	# Triple patterns	Query diameter
Q1	5	1
Q2	10	1
Q3a	2	1
Q3b	2	1
Q3c	2	1
Q4	8	2
Q5a	6	2
Q5b	6	2
Q6	9	2
Q7	13	5
Q8	8	2
Q9	4	2
Q10	1	1
Q11	1	1
Q12a	6	2
Q12b	8	2
Q12c	1	1

Load time is the time that the RDF store needs to parse and load the RDF graph.

Query execution time is the average time to execute a query.

Global query execution time is the arithmetic and geometric mean of all 17 queries. It is computed by multiplying the execution times of all 17 queries and then computing the 17th root of the result. If a query could not be processed, it is punished with 3600 s.

Memory consumption is measured by (i) the maximum amount of memory that was allocated during the processing of each individual query and (ii) the average memory consumption of all queries.

Berlin SPARQL Benchmark (BSBM). The BSBM[23] [21] focusses on the use case of an e-commerce platform. It aims to simulate the search and navigation patterns of multiple concurrently acting customers. The dataset is generated

---

[23] http://wifo5-03.informatik.uni-mannheim.de/bizer/berlinsparqlbenchmark/.

based on a relational schema. This schema represents products, their offers and
the custom reviews of the products. BSBM provides 12 query patterns whose
characteristics are given in Table 3. These query mainly test the ability to optimize the join ordering and the early application of filters. In BSBM a query
pattern refers to a query in which some constants are replaced by a special
type of variable. During the runtime of the benchmark these special variable
are replaced with varying constants occurring in the dataset. A set of queries in
which each query pattern is instantiated is called a query mix. Several of these
query mixes are executed in parallel in order to measure the performance of the
RDF store. The proposed performance metrics are:

Load time is the time that the RDF store needs to parse and load the RDF
   graph.
Query mixes per hour is the number of query mixes that can be completely
   processed within one hour.
Queries per second is the number of queries, which are instantiated from a single
   query pattern, that can be answered within one second.

Table 3. BSBM query characteristics.

Query	# Triple patterns	Query diameter
Q1	5	1
Q2	15	2
Q3	7	1
Q4	10	1
Q5	7	1
Q6	2	1
Q7	14	2
Q8	10	2
Q9	1	1
Q10	7	2
Q11	2	2
Q12	9	2

Semantic Publishing Benchmark (SPB). The SPB[24] [74] is a benchmark
motivated by the industry. The use case is a publisher organization that provides metadata about its published work. Many journalists search for data and
perform insertions and deletions concurrently. SPB provides a dataset generator
that is designed to create datasets with several billions of triples that mimic the

---

[24] http://ldbcouncil.org/developer/spb.

BBC datasets. Similar to BSBM it provides query templates that contains special variables that will be instantiated before query execution. SPB defines two set of query templates. The basic query set focuses on join ordering, duplicate elimination and filtering whereas the advanced query set additionally contains, e.g., analytical queries. The query characteristics are given in Table 4a and b. The proposed performance metrics are:

Minimum, maximum and average query execution time for each executed query. *Average execution rate per second* measures how many queries could be finished per second in average.

Table 4. SPB query characteristics.

Query	# Triple Patterns	Query Diameter
Q1	26	3
Q2	9	2
Q3	1	1
Q4	3	1
Q5	3	2
Q6	4	2
Q7	2	1
Q8	4	1
Q9	1	1
Q10	5	1
Q11	11	1
Q12	9	1
Q13	5	1
Q14	12	2
Q15	10	2
Q16	9	1
Q17	5	2
Q18	6	1
Q19	4	1
Q20	11	2
Q21	9	1
Q22	9	1
Q23	9	1
Q24	4	1
Q25	4	1

Query	# Triple Patterns	Query Diameter
Q1	26	3
Q2	9	2
Q3	8	1
Q4	4	1
Q5	6	2
Q6	5	2
Q7	6	1
Q8	11	2
Q9	9	1
Q10	7	3
Q11	8	2

(a) Basic query set characteristics.   (b) Advanced query set characteristics.

FedBench. FedBench [125] is designed as a benchmark for federated RDF stores. It provides three different dataset collections: (i) a general linked data collection containing DBPedia, GeoNames, Jamendo, Linked_MDB, New York Times and Semantic Web Dog Food, (ii) a life science data collection containing KEGG, ChEBI and DrugBank as well as (iii) a dataset of 10M triples generated with the dataset generator of SP²Bench. FedBench provides two self-made query sets as well as the queries from SP²Bench for the three data collections. The first two query sets focus on the number of data sources involved, the join ordering and query results set sizes. Since the actual benchmark cannot be found online any more, the characteristics of the queries cannot be examined. The proposed performance metrics are:

Query execution time for each executed query.
Number of requests to remote RDF stores during the processing of each query.

8.2   Benchmark Generators

DBPedia SPARQL Benchmark (DBSB). The general idea of DBSB [94] is to scale the DBPedia dataset to the required size and create queries based on a historic query log of DBPedia SPARQL endpoints. In order to generate the dataset a DBPedia dump is taken. To increase the size, triples are replicated and there namespaces are changed. To shrink the dataset size, triples are removed in a way that its characteristics like the indegree and the outdegree of vertices is not changed. In order to generate queries, DBSB clusters all queries of a historic query log. Out of each cluster the most frequent queries were picked as well as queries that cover most SPARQL features. Based on the selected queries, new queries are generated by replacing the constants with resources of the generated dataset during the benchmark generation process.

Waterloo SPARQL Diversity Test Suite (WatDiv). WatDiv [15] was designed to create benchmarks that are able to test the performance change of RDF stores under varying dataset and query characteristics. Therefore, the dataset generator is able to create datasets with variations of (a) the entity types, (b) the graph topology, (c) the well-structuredness of entities (i.e., which portion of the defined edges are usually present at an entity), (d) the probability of edges connecting two entities and (e) the cardinality of properties. In order to generate queries based on a dataset, the following characteristics are defined:

Triple Pattern Count defines the number of triple patterns occurring in the generated query.
Join Vertex Count counts the number of resources or variables that occur in multiple triple patterns.
Join Vertex Degree determined in how many triple patterns each join vertex occurs.
Join Vertex Type defines whether a subject-subject, subject-object or object-object join is performed.
Result Cardinality is the number of results.
Filter Triple Pattern Selectivity defines with witch portion of the graph a triple pattern matches.
BGP-Restricted f-TP Selectivity determines to which extent a triple pattern contributes to the overall selectivity of a query.
Join-Restricted f-TP Selectivity determines to which extent a triple pattern contributes to the overall selectivity of a join.

FEASIBLE. FEASIBLE [119] does not provide a dataset generator. Instead it can use an arbitrary dataset for which a historic query log exists. FEASIBLE aims to generate queries that have similar characteristics to the queries in the query log. Therefore, in a first step all syntactical incorrect queries and queries with no results are removed. Each query is transformed into a vector based on the following query characteristics:

SPARQL features defines which SPARQL features like SELECT, ASK, UNION, etc. occur in the query.

Triple Pattern Count defines the number of triple patterns occurring in the generated query.

Join Vertex Count counts the number of resources or variables that occur in multiple triple patterns.

Join Vertex Degree determined in how many triple patterns each join vertex occurs.

Join Vertex Type defines whether a subject-subject, subject-object or object-object join is performed.

Triple Pattern Selectivity defines with witch portion of the graph a triple pattern matches.

From the resulting set of query vectors, the requested number of queries are selected in a way that their vectors are as far away as possible from each other.

SPLODGE. The idea of SPLODGE [47] is to generated queries with a given set of characteristics from an arbitrary dataset. Thereby, it uses the following query characteristics:

Query Type defines whether a SELECT, CONSTRUCT, ASK or DESCRIBE query should be generated.

Join Type defines whether a conjunctive join (.), disjunctive join (UNION) or left-join (OPTIONAL) should be performed.

Result Modifiers defines whether the result set should be altered by DISTINCT, LIMIT, OFFSET or ORDER_BY operators.

Variable Patterns defines at which positions of the triple pattern variables should occur.

Join Patterns defines whether a subject-subject, subject-object or object-object join is performed.

Cross Products defines whether conjunctive join without join variable should be performed.

Number of Sources defines from how many different data sources triples should be combined to answer the query.

Number of Joins defines how many joins should occur in the query.

Query Selectivity defines with witch portion of the graph all triple patterns of a query match.

Table 5. Evaluations of RDF stores in the cloud published since 2016.

Paper	Benchmark	Max. dataset size	# compute nodes	compute node size
[29]	WatDiv	~100M triples	10	6 cores, 32 GB RAM, 4 TB disk
[112]	WatDiv	~1,000M triples	10	6 cores, 32 GB RAM, 4 TB disk
[96]	LUBM WatDiv	~1,330M triples	18	12 cores, 50 GB RAM
[92]	WatDiv	~10M triples	11	4 cores, 24 GB RAM
[86]	LUBM	~35M triples	18 slaves 1 federator	16 cores, 28 GB RAM 16 cores, 56 GB RAM
[5]	LUBM	~4,200M triples	12	24 cores, 148 GB RAM
[139]	LUBM	~220M triples	4–16 slaves 1 master	4 cores, 8 GB RAM, 500 GB disk 4 cores, 16 GB RAM, 500 GB disk
[124]	WatDiv	~1,000M triples	10	6 cores, 32 GB RAM, 4 TB disk
[121]	WatDiv	~100M triples	10	6 cores, 32 GB RAM, 4 TB disk
[109]	WatDiv LUBM	~1,382M triples	10	8 cores, 32 GB RAM
[106]	BSBM	~5M triples	4–12	2 cores, 8 GB RAM
[105]	WatDiv LUBM FedBench	~1,099M triples	10	4 cores, 16 GB RAM, 500 GB disk
[104]	WatDiv	~250M triples	10	4 cores, 16 GB RAM, 150 GB disk
[57]	WatDiv LUBM	~4,288M triples	5–12	24 cores, 148 GB RAM
[50]	WatDiv LUBM	~1,380M triples	10	4 cores, 17 GB RAM
[4]	LUBM	~3,100M triples	11	8 cores, 16 GB RAM, 3 TB disk

8.3   Performed Evaluations

The before mentioned benchmarks are usually used to evaluate and compare the performance of RDF stores as a whole. Table 5 summarizes the evaluations published from the beginning of 2016. All of them use are least one of the benchmarks described above. Beside the generated datasets they usually also use a few realistic datasets. The maximal dataset size is in most cases approximately 1 billion triples. In two cases a dataset with up to 4.2 billion triples was used. Most RDF stores in the cloud were deployed on 10 compute nodes. In one evaluation 19 compute nodes were used.

The evaluations in these papers use rather small datasets. To give a better overview of the capabilities of current RDF stores, [1] reports RDF stores running on a single server or in the cloud that could store RDF graphs consisting of several billions or even one trillion triples (see the summary in Table 6).

In order to compare the influence of alternative graph cover strategies or different query execution strategies, all but the examined component of the distributed RDF store need to stay the same. This was done, for instance in [96] to

Table 6. Evaluations of RDF stores reported by [1].

RDF store	Max. dataset size	# compute nodes	Compute node size
Oracle database 12c[a]	~1 trillion triples	1	360 cores, 2 TB RAM, 45 TB disk
AllegroGraph[b]	~1 trillion triples	1	?
Stardog[c]	~50,000M triples	1	32 cores, 256 GB RAM
Virtuoso Clustered Edition [41]	~37,000M triples	8	8 cores, 16 GB RAM, 4 TB disk
GraphDB[d]	~17,000M triples	1	16 cores, 512 GB RAM
4store [58]	~15,000M triples	9	?
Blazegraph [2]	~12,700M triples	?	?
YARS2 [2]	~7,000M triples	?	?
Jena TDB[e]	~1,700M triples	1	2 cores, 10 GB RAM
RDFox[f]	~1,700M triples	1	16 cores, 50 GB RAM

[a] http://www.oracle.com/us/corporate/features/database-12c/index.html.
[b] https://franz.com/agraph/allegrograph/.
[c] https://www.stardog.com/.
[d] www.ontotext.com/products/ontotext-graphdb/.
[e] https://jena.apache.org/.
[f] http://www.cs.ox.ac.uk/isg/tools/RDFox/.

compare different query execution strategies on top of Spark or in [33, 63, 79, 145] to compare different graph cover strategies. But these evaluation used Hadoop or its distributed file system to exchange data during the query processing. As a result, it remains unclear whether their results are applicable on distributed RDF stores in which the data is exchanged directly between the compute nodes.

9   Conclusion

To cope with the growing size of huge graphs, scalable RDF stores in the cloud are used, where the graph data is distributed among several compute nodes. From this distributed setting several challenges like (i) the data placement strategy, (ii) the distributed query processing, and (iii) the handling of failed compute nodes. In this manuscript we gave an overview of how these challenges are addressed by RDF stores in the cloud.

Due to the high number of RDF stores in the cloud, we have only given an overview of core challenges in distributed RDF stores. Beside these core challenges there exist further features that are required during the practical usage of relational databases in the industry today. Realizing them in RDF stores is a challenging task so that they are only partly realized in RDF stores. In order to achieve a broader usage of RDF stores in industry, further research is required to implement these features in RDF stores in the cloud. In the following we describe some of these features. As an example use case we assume an online retailer.

When two customers try to order a unique product at the same time, only one of the orders must be successful and the other must fail. To prevent the situation that both customers could successfully order the unique product, *transactional security* (i.e., atomicity, consistency, isolation and durability) is required. Realizing transactional security in a distributed setting where the data is separated among several compute nodes might require a lot of coordination between the compute nodes. This additional coordination increase the query execution time. To avoid this overhead while providing transactional security, most RDF stores in the cloud assume that the RDF graph is immutable after loading it. Only few RDF stores like Virtuoso Clustered Edition [40] allow for inserting or deleting triples after loading.

In order to identify which products were sold the most frequently in the last three month, the database is required to perform *online analytical processing (OLAP) queries*. This type of queries require a huge amount of data to be processed. In context of RDF stores in the cloud, processing OLAP queries cannot be done by sending all required data to a single compute node since a single compute node might be overloaded by the huge amount of data. Therefore, graph cover strategies and distributed query execution strategies need to be developed that support the parallel processing of OLAP queries with a low number of exchanged network packets.

To simplify the search for a product, the online retailer has categorized its products in a category hierarchy. For instance, an orange lemonade can be categorized as lemonade, soft drink and drink. With the introduction of SPARQL 1.1 [110] *property paths* were introduced that allow for requesting all offered drinks independently of the subcategory they belong to. This type of query differs from the pure graph pattern matching done in SPARQL 1.0 since it can easily require the traversal of long paths. In a distributed setting the traversal of long paths may lead to a high network traffic reducing the query execution time. One challenge arising from these queries is, how to optimize the data placement for these queries. Alternatively to SPARQL, the retailer might want to use other query languages like GraphQL[25].

The retailer wants to prevent teenagers from buying alcohol. Therefore, he stores in database the rules that every customer younger than 20 is a teenager and teenagers should not be allowed to buy alcohol. These rules should be automatically applied to all customers. In order to realize this, RDF stores need the ability to reason about RDF graphs. In this context *reasoning* means inferring logical consequences and checking the consistency of the RDF graph. Usually, reasoning is done during the loading time of the graph and all logical consequences are stored as explicit triples in the graph. The challenge of distributed reasoning is that the reasoning of the complete graph might overload a single compute node. Therefore, distributed reasoning algorithms are required. A few RDF stores in the cloud like MaRVIN [100] and Rya [114] have addressed this challenge. Another problem is, if the RDF graph is mutable after loading. In this case the deletion or insertion of triples might produce inconsistencies, a lot

---

[25] https://graphql.org/.

of newly inferred triples need to be inserted or formerly inferred triples need to be removed.

Finally, the retailer wants to advertise summer products like swimwear, portable fans, etc. more prominently if the temperature in the town where the customer lives is high. Therefore, the constantly streamed data from temperature sensors needs to be processed. This quickly arriving stream data cause further challenges for RDF stores, like quickly combining the received data with static data, updating the database frequently or balancing the workload among all compute nodes. CQELS Cloud [78] is one example of a system that processes RDF streams in a distributed fashion.

References

1. Largetriplestores. https://www.w3.org/wiki/LargeTripleStores. Accessed 10 July 2018
2. The bigdata® RDF Database. http://www.bigdata.com/whitepapers/bigdata_architecture_whitepaper.pdf. Accessed 29 Oct 2014
3. Abadi, D.J., Marcus, A., Madden, S.R., Hollenbach, K.: Scalable semantic web data management using vertical partitioning. In: Proceedings of the 33rd International Conference on Very Large Data Bases, VLDB 2007, pp. 411–422. VLDB Endowment (2007). http://dl.acm.org/citation.cfm?id=1325851.1325900
4. Abbassi, S., Faiz, R.: RDF-4X: a scalable solution for RDF quads store in the cloud. In: Proceedings of the 8th International Conference on Management of Digital EcoSystems, MEDES, pp. 231–236. ACM, New York (2016). https://doi.org/10.1145/3012071.3012104
5. Abdelaziz, I., Harbi, R., Salihoglu, S., Kalnis, P.: Combining vertex-centric graph processing with SPARQL for large-scale RDF data analytics. IEEE Trans. Parallel Distrib. Syst. **28**(12), 3374–3388 (2017). https://doi.org/10.1109/TPDS.2017.2720174
6. Aberer, K., Cudré-Mauroux, P., Hauswirth, M., Van Pelt, T.: GridVine: building internet-scale semantic overlay networks. In: McIlraith, S.A., Plexousakis, D., van Harmelen, F. (eds.) ISWC 2004. LNCS, vol. 3298, pp. 107–121. Springer, Heidelberg (2004). https://doi.org/10.1007/978-3-540-30475-3_9
7. Acosta, M., Vidal, M.-E., Lampo, T., Castillo, J., Ruckhaus, E.: ANAPSID: an adaptive query processing engine for SPARQL endpoints. In: Aroyo, L. (ed.) ISWC 2011. LNCS, vol. 7031, pp. 18–34. Springer, Heidelberg (2011). https://doi.org/10.1007/978-3-642-25073-6_2
8. Akar, Z., Halaç, T.G., Ekinci, E.E., Dikenelli, O.: Querying the web of interlinked datasets using VOID descriptions. In: WWW 2012 Workshop on Linked Data on the Web, Lyon, France, 16 April 2012. http://ceur-ws.org/Vol-937/ldow2012-paper-06.pdf
9. Al-Harbi, R., Abdelaziz, I., Kalnis, P., Mamoulis, N., Ebrahim, Y., Sahli, M.: Adaptive partitioning for very large RDF data. CoRR abs/1505.0 (2015). http://arxiv.org/abs/1505.02728
10. Al-Harbi, R., Ebrahim, Y., Kalnis, P.: PHD-store: an adaptive SPARQL engine with dynamic partitioning for distributed RDF repositories. CoRR abs/1405.4 (2014). http://arxiv.org/abs/1405.4979

11. Alexander, K., Cyganiak, R., Hausenblas, M., Zhao, J.: Describing linked datasets with the VoID vocabulary. W3C Interest Group Note, W3C (2011). http://www.w3.org/TR/2011/NOTE-void-20110303/

12. Ali, L., Janson, T., Lausen, G.: 3rdf: storing and querying RDF data on top of the 3nuts overlay network. In: 2011 22nd International Workshop on Database and Expert Systems Applications, pp. 257–261 (2011). https://doi.org/10.1109/DEXA.2011.1

13. Ali, L., Janson, T., Schindelhauer, C.: Towards load balancing and parallelizing of RDF query processing in P2P based distributed RDF data stores. In: 2014 22nd Euromicro International Conference on Parallel, Distributed, and Network-Based Processing, pp. 307–311 (2014). https://doi.org/10.1109/PDP.2014.79

14. Ali, L., Janson, T., Lausen, G., Schindelhauer, C.: Effects of network structure improvement on distributed RDF querying. In: Hameurlain, A., Rahayu, W., Taniar, D. (eds.) Globe 2013. LNCS, vol. 8059, pp. 63–74. Springer, Heidelberg (2013). https://doi.org/10.1007/978-3-642-40053-7_6

15. Aluç, G., Hartig, O., Özsu, M.T., Daudjee, K.: Diversified stress testing of RDF data management systems. In: Mika, P. (ed.) ISWC 2014. LNCS, vol. 8796, pp. 197–212. Springer, Cham (2014). https://doi.org/10.1007/978-3-319-11964-9_13

16. Arenas, M., Pérez, J.: Federation and navigation in SPARQL 1.1. In: Eiter, T., Krennwallner, T. (eds.) Reasoning Web 2012. LNCS, vol. 7487, pp. 78–111. Springer, Heidelberg (2012). https://doi.org/10.1007/978-3-642-33158-9_3

17. Basca, C., Bernstein, A.: Distributed SPARQL throughput increase: on the effectiveness of workload-driven RDF partitioning. In: ISWC 2013 (2013)

18. Basca, C., Bernstein, A.: Querying a messy web of data with AVALANCHE. Web Semant.: Sci. Serv. Agents World Wide Web **26** (2014). http://www.websemanticsjournal.org/index.php/ps/article/view/361

19. Battré, D., Heine, F., Höing, A., Kao, O.: On triple dissemination, forward-chaining, and load balancing in DHT based RDF stores. In: Moro, G., Bergamaschi, S., Joseph, S., Morin, J.-H., Ouksel, A.M. (eds.) DBISP2P 2005–2006. LNCS, vol. 4125, pp. 343–354. Springer, Heidelberg (2007). https://doi.org/10.1007/978-3-540-71661-7_33

20. Beame, P., Koutris, P., Suciu, D.: Skew in parallel query processing. In: Proceedings of the 33rd ACM SIGMOD-SIGACT-SIGART Symposium on Principles of Database Systems, PODS 2014, pp. 212–223. ACM, New York (2014). https://doi.org/10.1145/2594538.2594558

21. Bizer, C., Schultz, A.: The Berlin SPARQL benchmark. Int. J. Semant. Web Inf. Syst. **5**(2), 1–24 (2009). https://doi.org/10.4018/jswis.2009040101

22. Böhm, C., Hefenbrock, D., Naumann, F.: Scalable peer-to-peer-based RDF management. In: Proceedings of the 8th International Conference on Semantic Systems, I-SEMANTICS 2012, pp. 165–168. ACM, New York (2012). https://doi.org/10.1145/2362499.2362523

23. Bröcheler, M., Pugliese, A., Subrahmanian, V.S.: COSI: cloud oriented subgraph identification in massive social networks. In: Advances in Social Networks Analysis and Mining (ASONAM), pp. 248–255 (2010). https://doi.org/10.1109/ASONAM.2010.80

24. Bugiotti, F., Camacho-Rodríguez, J., Goasdoué, F., Kaoudi, Z., Manolescu, I., Zampetakis, S.: SPARQL query processing in the cloud. In: Harth, A., Hose, K., Schenkel, R. (eds.) Linked Data Management. Emerging Directions in Database Systems and Applications. Chapman and Hall/CRC (2014)

25. Cai, M., Frank, M.: RDFPeers: a scalable distributed RDF repository based on a structured peer-to-peer network. In: Proceedings of the 13th International Conference on World Wide Web, pp. 650–657 (2004). http://dl.acm.org/citation.cfm?id=988760

26. Charalambidis, A., Troumpoukis, A., Konstantopoulos, S.: SemaGrow: optimizing federated SPARQL queries. In: Proceedings of the 11th International Conference on Semantic Systems, SEMANTICS 2015, pp. 121–128. ACM, New York (2015). https://doi.org/10.1145/2814864.2814886

27. Cheng, L., Kotoulas, S.: Scale-out processing of large RDF datasets. IEEE Trans. Big Data 1(4), 138–150 (2015). https://doi.org/10.1109/TBDATA.2015.2505719

28. Chu, S., Balazinska, M., Suciu, D.: From theory to practice: efficient join query evaluation in a parallel database system. In: Proceedings of the 2015 ACM SIGMOD International Conference on Management of Data, SIGMOD 2015, pp. 63–78. ACM, New York (2015). https://doi.org/10.1145/2723372.2750545

29. Cossu, M., Färber, M., Lausen, G.: PRoST: distributed execution of SPARQL queries using mixed partitioning strategies. In: Proceedings of the 21st International Conference on Extending Database Technology, EDBT 2018, Vienna, Austria, 26–29 March 2018, pp. 469–472 (2018). https://doi.org/10.5441/002/edbt.2018.49

30. Crespo, A., Garcia-Molina, H.: Semantic overlay networks for P2P systems. In: Moro, G., Bergamaschi, S., Aberer, K. (eds.) AP2PC 2004. LNCS (LNAI), vol. 3601, pp. 1–13. Springer, Heidelberg (2005). https://doi.org/10.1007/11574781_1

31. Cudre-Mauroux, P., Agarwal, S., Aberer, K.: GridVine: an infrastructure for peer information management. IEEE Internet Comput. 11(5), 36–44 (2007). https://doi.org/10.1109/MIC.2007.108

32. Cudré-Mauroux, P., et al.: NoSQL databases for RDF: an empirical evaluation. In: Alani, H. (ed.) ISWC 2013. LNCS, vol. 8219, pp. 310–325. Springer, Heidelberg (2013). https://doi.org/10.1007/978-3-642-41338-4_20

33. Curé, O., Naacke, H., Baazizi, M.A., Amann, B.: On the evaluation of RDF distribution algorithms implemented over apache spark. In: Proceedings of the 11th International Workshop on Scalable Semantic Web Knowledge Base Systems (ISWC 2015), pp. 16–31 (2015)

34. DeCandia, G., et al.: Dynamo: Amazon's highly available key-value store. In: Proceedings of Twenty-first ACM SIGOPS Symposium on Operating Systems Principles, SOSP 2007, pp. 205–220. ACM, New York (2007). https://doi.org/10.1145/1294261.1294281

35. Della Valle, E., Turati, A., Ghioni, A.: *PAGE*: a distributed infrastructure for fostering rdf-based interoperability. In: Eliassen, F., Montresor, A. (eds.) DAIS 2006. LNCS, vol. 4025, pp. 347–353. Springer, Heidelberg (2006). https://doi.org/10.1007/11773887_27

36. DeWitt, D.J., Katz, R.H., Olken, F., Shapiro, L.D., Stonebraker, M.R., Wood, D.A.: Implementation techniques for main memory database systems. In: Proceedings of the 1984 ACM SIGMOD International Conference on Management of Data, SIGMOD 1984, pp. 1–8. ACM, New York (1984). https://doi.org/10.1145/602259.602261

37. Dhraief, H., Kemper, A., Nejdl, W., Wiesner, C.: Processing and optimization of complex queries in schema-based P2P-networks. In: Ng, W.S., Ooi, B.-C., Ouksel, A.M., Sartori, C. (eds.) DBISP2P 2004. LNCS, vol. 3367, pp. 31–45. Springer, Heidelberg (2005). https://doi.org/10.1007/978-3-540-31838-5_3

38. Ding, L., Peng, Y., da Silva, P.P., McGuinness, D.L.: Tracking RDF graph provenance using RDF molecules. Technical report, UMBC (2005). https://ebiquity. umbc.edu/paper/html/id/240/Tracking-RDF-Graph-Provenance-using-RDF-Molecules

39. Du, F., Bian, H., Chen, Y., Du, X.: Efficient SPARQL query evaluation in a database cluster. In: IEEE International Congress on Big Data, pp. 165–172 (2013). https://doi.org/10.1109/BigData.Congress.2013.30

40. Erling, O., Mikhailov, I.: Towards web scale RDF. In: 4th International Workshop on Scalable Semantic Web Knowledge Base Systems (SSWS 2008) (2008)

41. Erling, O., Mikhailov, I.: Virtuoso: RDF support in a native RDBMS. In: de Virgilio, R., Giunchiglia, F., Tanca, L. (eds.) Semantic Web Information Management, pp. 501–519. Springer, Heidelberg (2010). https://doi.org/10.1007/978-3-642-04329-1_21

42. Farhan Husain, M., McGlothlin, J., Masud, M.M., Khan, L., Thuraisingham, B.: Heuristics-based query processing for large RDF graphs using cloud computing. IEEE Trans. Knowl. Data Eng. **23**(9), 1312–1327 (2011). https://doi.org/10.1109/TKDE.2011.103

43. Galarraga, L., Hose, K., Schenkel, R.: Partout: a distributed engine for efficient RDF processing. CoRR abs/1212.5 (2012). http://arxiv.org/abs/1212.5636

44. Goasdoué, F., Kaoudi, Z., Manolescu, I., Quiané-Ruiz, J.A., Zampetakis, S.: CliqueSquare: flat plans for massively parallel RDF queries. In: 2015 IEEE 31st International Conference on Data Engineering, pp. 771–782 (2015). https://doi.org/10.1109/ICDE.2015.7113332

45. Gonzalez, J.E., Xin, R.S., Dave, A., Crankshaw, D., Franklin, M.J., Stoica, I.: GraphX: graph processing in a distributed dataflow framework. In: Proceedings of the 11th USENIX Conference on Operating Systems Design and Implementation, OSDI 2014, pp. 599–613. USENIX Association, Berkeley (2014). http://dl.acm.org/citation.cfm?id=2685048.2685096

46. Goodman, E.L., Grunwald, D.: Using vertex-centric programming platforms to implement SPARQL queries on large graphs. In: Proceedings of the 4th Workshop on Irregular Applications: Architectures and Algorithms, IA³ 2014, pp. 25–32. IEEE Press, Piscataway (2014). https://doi.org/10.1109/IA3.2014.10

47. Görlitz, O., Thimm, M., Staab, S.: SPLODGE: Systematic generation of SPARQL benchmark queries for linked open data. Semant. Web-ISWC **2012**, 116–132 (2012). https://doi.org/10.1007/978-3-642-35176-1_8

48. Görlitz, O., Staab, S.: SPLENDID: SPARQL endpoint federation exploiting VOID descriptions. In: Proceedings of the Second International Conference on Consuming Linked Data, COLD 2011, vol. 782, pp. 13–24. CEUR-WS.org, Aachen (2010). http://dl.acm.org/citation.cfm?id=2887352.2887354

49. Graux, D., Jachiet, L., Genevès, P., Layaïda, N.: A Multi-Criteria Experimental Ranking of Distributed SPARQL Evaluators (2016). https://hal.inria.fr/hal-01381781

50. Graux, D., Jachiet, L., Genevès, P., Layaïda, N.: SPARQLGX: efficient distributed evaluation of SPARQL with apache spark. In: Groth, P., et al. (eds.) ISWC 2016. LNCS, vol. 9982, pp. 80–87. Springer, Cham (2016). https://doi.org/10.1007/978-3-319-46547-0_9

51. Guo, Y., Pan, Z., Heflin, J.: LUBM: a benchmark for owl knowledge base systems. Web Semant.: Sci. Serv. Agents World Wide Web, **3**(2–3) (2005). http://www.websemanticsjournal.org/index.php/ps/article/view/70

52. Gurajada, S., Seufert, S., Miliaraki, I., Theobald, M.: TriAD: a distributed shared-nothing RDF engine based on asynchronous message passing. In: SIGMOD, pp. 289–300 (2014). https://doi.org/10.1145/2588555.2610511

53. Gutierrez, C., Hurtado, C., Mendelzon, A.O.: Foundations of semantic web databases. In: PODS, pp. 95–106. ACM (2004). https://doi.org/10.1145/1055558. 1055573

54. Haas, L.M., Kossmann, D., Wimmers, E.L., Yang, J.: Optimizing queries across diverse data sources. In: VLDB 1997, Athens, Greece, pp. 276–285. Morgan Kaufmann Publishers Inc., San Francisco (1997)

55. Hammoud, M., Rabbou, D.A., Nouri, R., Beheshti, S.M.R., Sakr, S.: DREAM: distributed RDF engine with adaptive query planner and minimal communication. Proc. VLDB Endow. 8(6), 654–665 (2015). https://doi.org/10.14778/2735703. 2735705

56. Harbi, R., Abdelaziz, I., Kalnis, P., Mamoulis, N.: Evaluating SPARQL queries on massive RDF datasets. PVLDB, 8(12), 1848–1851 (2015). http://www.vldb. org/pvldb/vol8/p1848-harbi.pdf

57. Harbi, R., Abdelaziz, I., Kalnis, P., Mamoulis, N., Ebrahim, Y., Sahli, M.: Accelerating SPARQL queries by exploiting hash-based locality and adaptive partitioning. VLDB J. 25(3), 355–380 (2016). https://doi.org/10.1007/s00778-016-0420-y

58. Harris, S., Lamb, N., Shadbolt, N.: 4store: the design and implementation of a clustered RDF store. In: Scalable Semantic Web Knowledge Base Systems - SSWS 2009, pp. 94–109 (2009)

59. Harth, A., Decker, S.: Optimized index structures for querying RDF from the web. In: Proceedings of LA-WEB 2005, p. 71. IEEE (2005). https://doi.org/10. 1109/LAWEB.2005.25

60. Harth, A., Umbrich, J., Hogan, A., Decker, S.: YARS2: a federated repository for querying graph structured data from the web. In: Aberer, K. (ed.) ASWC/ISWC -2007. LNCS, vol. 4825, pp. 211–224. Springer, Heidelberg (2007). https://doi. org/10.1007/978-3-540-76298-0_16

61. Hong, S., Depner, S., Manhardt, T., Van Der Lugt, J., Verstraaten, M., Chafi, H.: PGX.D: a fast distributed graph processing engine. In: Proceedings of the International Conference for High Performance Computing, Networking, Storage and Analysis, SC 2015, pp. 58:1–58:12. ACM, New York (2015). https://doi.org/ 10.1145/2807591.2807620

62. Hose, K., Schenkel, R.: WARP: workload-aware replication and partitioning for RDF. In: Data Engineering Workshops (ICDEW), pp. 1–6 (2013). https://doi. org/10.1109/ICDEW.2013.6547414

63. Huang, J., Abadi, D.J., Ren, K.: Scalable SPARQL querying of large RDF graphs. PVLDB 4(11), 1123–1134 (2011)

64. Janke, D., Staab, S., Thimm, M.: Impact analysis of data placement strategies on query efforts in distributed RDF stores. J. Web Semant. (2018). https://doi.org/ 10.1016/j.websem.2018.02.002, http://www.websemanticsjournal.org/index.php/ ps/article/view/516

65. Jones, N.D.: An introduction to partial evaluation. ACM Comput. Surv. 28(3), 480–503 (1996). https://doi.org/10.1145/243439.243447

66. Kang, U., Tsourakakis, C.E., Faloutsos, C.: PEGASUS: a peta-scale graph mining system implementation and observations. In: 2009 Ninth IEEE International Conference on Data Mining, pp. 229–238 (2009). https://doi.org/10.1109/ICDM. 2009.14

67. Kaoudi, Z., Koubarakis, M., Kyzirakos, K., Miliaraki, I., Magiridou, M., Papadakis-Pesaresi, A.: Atlas: storing, updating and querying RDF(S) data on top of DHTs. Web Semant.: Sci. Serv. Agents World Wide Web **8**(4) (2010). http://www.websemanticsjournal.org/index.php/ps/article/view/250

68. Karnstedt, M., et al.: UniStore: querying a DHT-based universal storage. In: 2007 IEEE 23rd International Conference on Data Engineering, pp. 1503–1504 (2007). https://doi.org/10.1109/ICDE.2007.369054

69. Karypis, G., Kumar, V.: A fast and high quality multilevel scheme for partitioning irregular graphs. SIAM J. Sci. Comput. **20**(1), 359–392 (1998). https://doi.org/10.1137/S1064827595287997

70. Khadilkar, V., Kantarcioglu, M., Thuraisingham, B.M., Castagna, P.: Jena-HBase: a distributed, scalable and efficient RDF triple store. Technical report, Department of Computer Science at the University of Texas at Dallas (2012)

71. Khadilkar, V., Kantarcioglu, M., Thuraisingham, B.M., Castagna, P.: Jena-HBase: a distributed, scalable and efficient RDF triple store. In: Proceedings of the ISWC 2012 Posters & Demonstrations Track, Boston, USA, 11–15 November 2012. http://ceur-ws.org/Vol-914/paper_14.pdf

72. Kim, H., Ravindra, P., Anyanwu, K.: From SPARQL to MapReduce: the journey using a nested TripleGroup Algebra. PVLDB **4**(12), 1426–1429 (2011). http://www.vldb.org/pvldb/vol4/p1426-kim.pdf

73. Kokkinidis, G., Christophides, V.: Semantic query routing and processing in P2P database systems: the ICS-FORTH SQPeer middleware. In: Lindner, W., Mesiti, M., Türker, C., Tzitzikas, Y., Vakali, A.I. (eds.) EDBT 2004. LNCS, vol. 3268, pp. 486–495. Springer, Heidelberg (2005). https://doi.org/10.1007/978-3-540-30192-9_48

74. Kotsev, V., Kiryakov, A., Fundulaki, I., Alexiev, V.: LDBC semantic publishing benchmark (SPB) - v2.0 first public draft release. Technical report, The Linked Data Benchmark Council (2014). https://github.com/ldbc/ldbc_spb_bm_2.0/blob/master/doc/LDBC_SPB_v2.0.docx?raw=true

75. Ladwig, G., Harth, A.: CumulusRDF: linked data management on nested key-value stores. In: Proceedings of the 7th International Workshop on Scalable Semantic Web Knowledge Base Systems (SSWS 2011) at the 10th International Semantic Web Conference (ISWC 2011) (2011)

76. Ladwig, G., Tran, T.: SIHJoin: querying remote and local linked data. In: Antoniou, G., et al. (eds.) ESWC 2011. LNCS, vol. 6643, pp. 139–153. Springer, Heidelberg (2011). https://doi.org/10.1007/978-3-642-21034-1_10

77. Lakshman, A., Malik, P.: Cassandra: a decentralized structured storage system. SIGOPS Oper. Syst. Rev. **44**(2), 35–40 (2010). https://doi.org/10.1145/1773912.1773922

78. Le-Phuoc, D., Nguyen Mau Quoc, H., Le Van, C., Hauswirth, M.: Elastic and scalable processing of linked stream data in the cloud. In: Alani, H., et al. (eds.) ISWC 2013. LNCS, vol. 8218, pp. 280–297. Springer, Heidelberg (2013). https://doi.org/10.1007/978-3-642-41335-3_18

79. Lee, K., Liu, L.: Efficient data partitioning model for heterogeneous graphs in the cloud. In: Proceedings of the International Conference on High Performance Computing, Networking, Storage and Analysis. pp. 46:1–46:12. ACM (2013). https://doi.org/10.1145/2503210.2503302

80. Lee, K., Liu, L.: Scaling queries over Big RDF graphs with semantic hash partitioning. PVLDB **6**(14), 1894–1905 (2013). https://doi.org/10.14778/2556549.2556571

81. Lee, K., Liu, L., Tang, Y., Zhang, Q., Zhou, Y.: Efficient and customizable data partitioning framework for distributed big RDF data processing in the cloud. In: IEEE CLOUD 2013, pp. 327–334 (2013). https://doi.org/10.1109/CLOUD.2013.63

82. Liarou, E., Idreos, S., Koubarakis, M.: Evaluating conjunctive triple pattern queries over large structured overlay networks. In: Cruz, I., et al. (eds.) ISWC 2006. LNCS, vol. 4273, pp. 399–413. Springer, Heidelberg (2006). https://doi.org/10.1007/11926078_29

83. Lynden, S., Kojima, I., Matono, A., Tanimura, Y.: ADERIS: an adaptive query processor for joining federated SPARQL endpoints. In: Meersman, R. (ed.) OTM 2011. LNCS, vol. 7045, pp. 808–817. Springer, Heidelberg (2011). https://doi.org/10.1007/978-3-642-25106-1_28

84. Malewicz, G., et al.: Pregel: a system for large-scale graph processing. In: Proceedings of the 2010 ACM SIGMOD International Conference on Management of Data, SIGMOD 2010, pp. 135–146. ACM, New York (2010). https://doi.org/10.1145/1807167.1807184

85. Malliaros, F.D., Vazirgiannis, M.: Clustering and community detection in directed networks: a survey. Phys. Rep. **533**(4), 95–142 (2013). https://doi.org/10.1016/j.physrep.2013.08.002

86. Mansour, E., Abdelaziz, I., Ouzzani, M., Aboulnaga, A., Kalnis, P.: A demonstration of Lusail: querying linked data at scale. In: Proceedings of the 2017 ACM International Conference on Management of Data, SIGMOD 2017, pp. 1603–1606. ACM, New York (2017). https://doi.org/10.1145/3035918.3058731

87. Matono, A., Pahlevi, S.M., Kojima, I.: RDFCube: a P2P-based three-dimensional index for structural joins on distributed triple stores. In: Moro, G., Bergamaschi, S., Joseph, S., Morin, J.-H., Ouksel, A.M. (eds.) DBISP2P 2005-2006. LNCS, vol. 4125, pp. 323–330. Springer, Heidelberg (2007). https://doi.org/10.1007/978-3-540-71661-7_31

88. McMurry, J., et al.: Report on the scalability of semantic web integration in biomedbridges (2015). https://doi.org/10.5281/zenodo.14071

89. Mishra, P., Eich, M.H.: Join processing in relational databases. ACM Comput. Surv. **24**(1), 63–113 (1992). https://doi.org/10.1145/128762.128764

90. Montoya, G., Skaf-Molli, H., Hose, K.: The *Odyssey* approach for optimizing federated SPARQL queries. In: d'Amato, C., et al. (eds.) ISWC 2017. LNCS, vol. 10587, pp. 471–489. Springer, Cham (2017). https://doi.org/10.1007/978-3-319-68288-4_28

91. Montoya, G., Skaf-Molli, H., Molli, P., Vidal, M.-E.: Federated SPARQL queries processing with replicated fragments. In: Arenas, M., et al. (eds.) ISWC 2015. LNCS, vol. 9366, pp. 36–51. Springer, Cham (2015). https://doi.org/10.1007/978-3-319-25007-6_3

92. Montoya, G., Skaf-Molli, H., Molli, P., Vidal, M.E.: Decomposing federated queries in presence of replicated fragments. Web Semant.: Sci. Serv. Agents World Wide Web **42**(1) (2017). http://www.websemanticsjournal.org/index.php/ps/article/view/486

93. Montoya, G., Vidal, M.E., Acosta, M.: A heuristic-based approach for planning federated SPARQL queries. In: Proceedings of the Third International Conference on Consuming Linked Data, COLD 2012, vol. 905, pp. 63–74. CEUR-WS.org, Aachen (2012). http://dl.acm.org/citation.cfm?id=2887367.2887373

94. Morsey, M., Lehmann, J., Auer, S., Ngonga Ngomo, A.-C.: DBpedia SPARQL benchmark – performance assessment with real queries on real data. In: Aroyo,

L., et al. (eds.) ISWC 2011. LNCS, vol. 7031, pp. 454–469. Springer, Heidelberg (2011). https://doi.org/10.1007/978-3-642-25073-6_29

95. Mutharaju, R., Sakr, S., Sala, A., Hitzler, P.: D-SPARQ: distributed, scalable and efficient RDF query engine. In: ISWC (Posters & Demos) 2013, pp. 261–264 (2013)

96. Naacke, H., Amann, B., Curé, O.: SPARQL graph pattern processing with apache spark. In: Proceedings of the Fifth International Workshop on Graph Data-Management Experiences and Systems, GRADES 2017, pp. 1:1–1:7. ACM, New York (2017). https://doi.org/10.1145/3078447.3078448

97. Nejdl, W., et al.: Super-peer-based routing and clustering strategies for RDF-based peer-to-peer networks. In: Proceedings of the 12th International Conference on World Wide Web, WWW 2003, pp. 536–543. ACM, New York (2003). https://doi.org/10.1145/775152.775229

98. Norvig, P.: The semantic web and the semantics of the web: where does meaning come from? In: Proceedings of the 25th International Conference on World Wide Web, WWW 2016, p. 1. International World Wide Web Conferences Steering Committee, Republic and Canton of Geneva (2016)

99. Olston, C., Reed, B., Srivastava, U., Kumar, R., Tomkins, A.: Pig Latin: a not-so-foreign language for data processing. In: Proceedings of the 2008 ACM SIGMOD International Conference on Management of Data, SIGMOD 2008, pp. 1099–1110. ACM, New York (2008). https://doi.org/10.1145/1376616.1376726

100. Oren, E., Kotoulas, S., Anadiotis, G., Siebes, R., ten Teije, A., van Harmelen, F.: Marvin: distributed reasoning over large-scale Semantic Web data. Web Semant.: Sci. Serv. Agents World Wide Web 7(4) (2009). http://www.websemanticsjournal.org/index.php/ps/article/view/173

101. Osorio, M., Aranda, C.B.: Storage balancing in P2P based distributed RDF data stores. In: Proceedings of the Workshop on Decentralizing the Semantic Web 2017, Co-located with 16th International Semantic Web Conference (ISWC 2017) (2017). http://ceur-ws.org/Vol-1934/contribution-04.pdf

102. Owens, A., Seaborne, A., Gibbins, N., schraefel, M.: Clustered TDB: A Clustered Triple Store for Jena (2008). http://eprints.soton.ac.uk/266974/

103. Papailiou, N., Tsoumakos, D., Konstantinou, I., Karras, P., Koziris, N.: H2RDF+: an efficient data management system for big RDF graphs. In: Proceedings of the 2014 ACM SIGMOD International Conference on Management of Data, SIGMOD 2014, pp. 909–912. ACM, New York (2014). https://doi.org/10.1145/2588555.2594535

104. Peng, P., Zou, L., Chen, L., Zhao, D.: Query workload-based RDF graph fragmentation and allocation. In: Proceedings of the 19th International Conference on Extending Database Technology, EDBT 2016, Bordeaux, France, 15–16 March 2016, pp. 377–388 (2016). https://doi.org/10.5441/002/edbt.2016.35

105. Peng, P., Zou, L., Özsu, M.T., Chen, L., Zhao, D.: Processing SPARQL queries over distributed RDF graphs. VLDB J. 25(2), 243–268 (2016). https://doi.org/10.1007/s00778-015-0415-0

106. Penteado, R.R.M., Scroeder, R., Hara, C.S.: Exploring controlled RDF distribution. In: 2016 IEEE International Conference on Cloud Computing Technology and Science (CloudCom), pp. 160–167 (2016). https://doi.org/10.1109/CloudCom.2016.0038

107. Pérez, J., Arenas, M., Gutierrez, C.: Semantics and complexity of SPARQL. ACM Trans. Database Syst. 34(3), 16:1–16:45 (2009). https://doi.org/10.1145/1567274.1567278

108. Potter, A., Motik, B., Horrocks, I.: Querying distributed RDF graphs: the effects of partitioning. In: Workshop on Scalable Semantic Web Knowledge Base Systems (SSWS 2014), pp. 29–44 (2014)
109. Potter, A., Motik, B., Nenov, Y., Horrocks, I.: Distributed RDF query answering with dynamic data exchange. In: Groth, P., et al. (eds.) ISWC 2016. LNCS, vol. 9981, pp. 480–497. Springer, Cham (2016). https://doi.org/10.1007/978-3-319-46523-4_29
110. Prud'hommeaux, E., Harris, S., Seaborne, A.: SPARQL 1.1 Query Language. W3C Recommendation, W3C (2013). http://www.w3.org/TR/sparql11-query/
111. Przyjaciel-Zablocki, M., Schätzle, A., Lausen, G.: TriAL-QL: distributed processing of navigational queries. In: Proceedings of the 18th International Workshop on Web and Databases, WebDB 2015, pp. 48–54, ACM, New York (2015). https://doi.org/10.1145/2767109.2767115
112. Przyjaciel-Zablocki, M., Schätzle, A., Lausen, G.: Querying semantic knowledge bases with SQL-on-Hadoop. In: Proceedings of the 4th ACM SIGMOD Workshop on Algorithms and Systems for MapReduce and Beyond, BeyondMR 2017, pp. 4:1–4:10. ACM, New York (2017). https://doi.org/10.1145/3070607.3070610
113. Pujol, J.M., Erramilli, V., Rodriguez, P.: Divide and conquer: partitioning online social networks. CoRR abs/0905.4 (2009). http://arxiv.org/abs/0905.4918
114. Punnoose, R., Crainiceanu, A., Rapp, D.: Rya: a scalable RDF triple store for the clouds. In: 1st International Workshop on Cloud Intelligence, pp. 4:1–4:8. ACM (2012). https://doi.org/10.1145/2347673.2347677
115. Quilitz, B., Leser, U.: Querying distributed RDF data sources with SPARQL. In: Bechhofer, S., Hauswirth, M., Hoffmann, J., Koubarakis, M. (eds.) ESWC 2008. LNCS, vol. 5021, pp. 524–538. Springer, Heidelberg (2008). https://doi.org/10.1007/978-3-540-68234-9_39
116. Rohloff, K., Schantz, R.E.: High-performance, massively scalable distributed systems using the MapReduce software framework: the SHARD triple-store. In: Programming Support Innovations for Emerging Distributed Applications, PSI EtA 2010, pp. 4:1–4:5. ACM, New York (2010). https://doi.org/10.1145/1940747.1940751
117. Russell, J.: Getting Started with Impala: Interactive SQL for Apache Hadoop. O'Reilly Media (2014). http://shop.oreilly.com/product/0636920033936.do
118. Sakr, S., Wylot, M., Mutharaju, R., Le Phuoc, D., Fundulaki, I., I.: Linked Data: Storing, Querying, and Reasoning. Springer, Cham (2018). https://doi.org/10.1007/978-3-319-73515-3
119. Saleem, M., Mehmood, Q., Ngonga Ngomo, A.-C.: FEASIBLE: a feature-based SPARQL benchmark generation framework. In: Arenas, M. (ed.) ISWC 2015. LNCS, vol. 9366, pp. 52–69. Springer, Cham (2015). https://doi.org/10.1007/978-3-319-25007-6_4
120. Saleem, M., Ngonga Ngomo, A.-C., Xavier Parreira, J., Deus, H.F., Hauswirth, M.: DAW: duplicate-aware federated query processing over the web of data. In: Alani, H., et al. (eds.) ISWC 2013. LNCS, vol. 8218, pp. 574–590. Springer, Heidelberg (2013). https://doi.org/10.1007/978-3-642-41335-3_36
121. Schätzle, A., Przyjaciel-Zablocki, M., Berberich, T., Lausen, G.: S2X: graph-parallel querying of RDF with GraphX. In: Wang, F., Luo, G., Weng, C., Khan, A., Mitra, P., Yu, C. (eds.) Big-O(Q)/DMAH -2015. LNCS, vol. 9579, pp. 155–168. Springer, Cham (2016). https://doi.org/10.1007/978-3-319-41576-5_12
122. Schätzle, A., Przyjaciel-Zablocki, M., Lausen, G.: PigSPARQL: mapping SPARQL to pig Latin. In: Proceedings of the International Workshop on Semantic

Web Information Management, SWIM 2011, pp. 4:1–4:8. ACM, New York (2011). https://doi.org/10.1145/1999299.1999303

123. Schätzle, A., Przyjaciel-Zablocki, M., Neu, A., Lausen, G.: Sempala: interactive SPARQL query processing on hadoop. In: Mika, P., et al. (eds.) ISWC 2014. LNCS, vol. 8796, pp. 164–179. Springer, Cham (2014). https://doi.org/10.1007/978-3-319-11964-9_11

124. Schätzle, A., Przyjaciel-Zablocki, M., Skilevic, S., Lausen, G.: S2RDF: RDF querying with SPARQL on spark. PVLDB 9(10), 804–815 (2016). http://www.vldb.org/pvldb/vol9/p804-schaetzle.pdf

125. Schmidt, M., Görlitz, O., Haase, P., Ladwig, G., Schwarte, A., Tran, T.: FedBench: a benchmark suite for federated semantic data query processing. In: Aroyo, L. (ed.) ISWC 2011. LNCS, vol. 7031, pp. 585–600. Springer, Heidelberg (2011). https://doi.org/10.1007/978-3-642-25073-6_37

126. Schmidt, M., Hornung, T., Meier, M., Pinkel, C., Lausen, G.: SP²Bench: a SPARQL performance benchmark. In: de Virgilio, R., Giunchiglia, F., Tanca, L. (eds.) Semantic Web Information Management: A Model-Based Perspective, pp. 371–393. Springer, Heidelberg (2010). https://doi.org/10.1007/978-3-642-04329-1_16

127. Schwarte, A., Haase, P., Hose, K., Schenkel, R., Schmidt, M.: FedX: optimization techniques for federated query processing on linked data. In: Aroyo, L., et al. (eds.) ISWC 2011. LNCS, vol. 7031, pp. 601–616. Springer, Heidelberg (2011). https://doi.org/10.1007/978-3-642-25073-6_38

128. Shvachko, K., Kuang, H., Radia, S., Chansler, R.: The Hadoop distributed file system. In: 2010 IEEE 26th Symposium on Mass Storage Systems and Technologies (MSST), pp. 1–10 (2010). https://doi.org/10.1109/MSST.2010.5496972

129. Stein, R., Zacharias, V.: RDF on cloud number nine. In: Ceri, S., Valle, E.D., Hendler, J., Huang, Z. (eds.) Proceedings of the 4th Workshop on New Forms of Reasoning for the Semantic Web: Scalable & Dynamic. CEUR Workshop Proceedings (2010)

130. Stutz, P., Verman, M., Fischer, L., Bernstein, A.: TripleRush: a fast and scalable triple store. In: 9th International Workshop on Scalable Semantic Web Knowledge Base Systems. CEUR Workshop Proceedings, Aachen (2013). http://ceur-ws.org

131. Stutz, P., Bernstein, A., Cohen, W.: Signal/collect: graph algorithms for the (semantic) web. ISWC 2010. LNCS, vol. 6496, pp. 764–780. Springer, Heidelberg (2010). https://doi.org/10.1007/978-3-642-17746-0_48

132. Stutz, P., Paudel, B., Verman, M., Bernstein, A.: Random walk TripleRush: asynchronous graph querying and sampling. In: Proceedings of the 24th International Conference on World Wide Web, WWW 2015, pp. 1034–1044. International World Wide Web Conferences Steering Committee, Republic and Canton of Geneva (2015). https://doi.org/10.1145/2736277.2741687

133. Wang, R., Chiu, K.: Optimizing distributed RDF triplestores via a locally indexed graph partitioning. In: 2012 41st International Conference on Parallel Processing (ICPP), pp. 259–268 (2012). https://doi.org/10.1109/ICPP.2012.47

134. Wang, X., Tiropanis, T., Davis, H.C.: LHD: optimising linked data query processing using parallelisation. In: Proceedings of the WWW 2013 Workshop on Linked Data on the Web, Rio de Janeiro, Brazil, 14 May 2013. http://ceur-ws.org/Vol-996/papers/ldow2013-paper-06.pdf

135. White, T.: Hadoop: The Definitive Guide, 4th edn. O'Reilly, Beijing (2015). https://www.safaribooksonline.com/library/view/hadoop-the-definitive/9781491901687/

136. Wilschut, A.N., Apers, P.M.G.: Dataflow query execution in a parallel main-memory environment. Distrib. Parallel Databases **1**(1), 103–128 (1993). https://doi.org/10.1007/BF01277522

137. Wu, B., Zhou, Y., Yuan, P., Liu, L., Jin, H.: Scalable SPARQL querying using path partitioning. In: 2015 IEEE 31st International Conference on Data Engineering, pp. 795–806 (2015). https://doi.org/10.1109/ICDE.2015.7113334

138. Wu, B., Zhou, Y., Yuan, P., Jin, H., Liu, L.: SemStore: a semantic-preserving distributed RDF triple store. In: CIKM 2014 (2014)

139. Wylot, M., Cudré-Mauroux, P.: Diplocloud: efficient and scalable management of rdf data in the cloud. IEEE Trans. Knowl. Data Eng. **28**(3), 659–674 (2016). https://doi.org/10.1109/TKDE.2015.2499202

140. Xu, Z., Chen, W., Gai, L., Wang, T.: SparkRDF: in-memory distributed RDF management framework for large-scale social data. In: Dong, X.L., Yu, X., Li, J., Sun, Y. (eds.) WAIM 2015. LNCS, vol. 9098, pp. 337–349. Springer, Cham (2015). https://doi.org/10.1007/978-3-319-21042-1_27

141. Yang, S., Yan, X., Zong, B., Khan, A.: Towards effective partition management for large graphs. In: Proceedings of the 2012 ACM SIGMOD International Conference on Management of Data, SIGMOD 2012, pp. 517–528. ACM, New York (2012). https://doi.org/10.1145/2213836.2213895

142. Yang, T., Chen, J., Wang, X., Chen, Y., Du, X.: Efficient SPARQL query evaluation via automatic data partitioning. In: Meng, W., Feng, L., Bressan, S., Winiwarter, W., Song, W. (eds.) DASFAA 2013. LNCS, vol. 7826, pp. 244–258. Springer, Heidelberg (2013). https://doi.org/10.1007/978-3-642-37450-0_18

143. Zaharia, M., Chowdhury, M., Franklin, M.J., Shenker, S., Stoica, I.: Spark: cluster computing with working sets. In: Proceedings of the 2nd USENIX Conference on Hot Topics in Cloud Computing, HotCloud 2010, p. 10. USENIX Association, Berkeley (2010). http://dl.acm.org/citation.cfm?id=1863103.1863113

144. Zeng, K., Yang, J., Wang, H., Shao, B., Wang, Z.: A distributed graph engine for web scale RDF data. PVLDB **6**(4), 265–276 (2013). https://doi.org/10.14778/2535570.2488333

145. Zhang, X., Chen, L., Tong, Y., Wang, M.: EAGRE: towards scalable I/O efficient SPARQL query evaluation on the cloud. In: ICDE 2013, pp. 565–576 (2013). https://doi.org/10.1109/ICDE.2013.6544856

146. Zhang, X., Chen, L., Wang, M.: Towards efficient join processing over large RDF graph using MapReduce. In: Ailamaki, A., Bowers, S. (eds.) SSDBM 2012. LNCS, vol. 7338, pp. 250–259. Springer, Heidelberg (2012). https://doi.org/10.1007/978-3-642-31235-9_16

Engineering of Web Stream Processing Applications

Emanuele Della Valle(✉), Riccardo Tommasini, and Marco Balduini

Department of Electronic, Informatics and Bioengineering, Politecnico di Milano,
Piazza Leonardo Da Vinci 32, 20191 Milan, Italy
{emanuele.dellavalle,riccardo.tommasini,marco.balduini}@polimi.it

Abstract. The goal of the tutorial is to outline how to develop and deploy a stream processing application in a Web environment in a reproducible way. To this extent, we intend to (1) survey existing research outcomes from the Stream Reasoning/RDF Stream Processing that arise in querying and reasoning on a variety of highly dynamic data, (2) introduce stream reasoning techniques as powerful tools to use when addressing a data-centric problem characterized both by variety and velocity (such as those typically found on the modern Web), (3) present a relevant Web-centric use-case that requires to address simultaneously data velocity and variety, and (4) guide the participants through the development of a Web stream processing application.

1 Introduction

More and more streams of information are becoming available on the Web. A variety of sources give origin to data streams including social networks, mobile phones, smart homes, healthcare devices and other modern infrastructures. A significant portion of this data belongs to the Web-of-Services and to the Web-of-Things ecosystems where data is published and consumed using Web standards and technologies.

From new opportunities arise new challenges. A common problem in the scenarios illustrated above is how to integrate such data and how to enable the creation of new knowledge. Reasoning techniques are a possible solution. However, while reasoners scale up in the classical, static domain of ontological knowledge, reasoning upon rapidly changing information has received attention only in the last decade [1]. The combination of reasoning techniques with data streams gives rise to Stream Reasoning. i.e., reasoning on highly dynamic flows of informations [2]. This is a high impact research area that has already started to produce results relevant to both the Semantic Web [3] and stream processing communities [4].

Learning Goals: The contents of this tutorial can be relevant for RW 2018 summer school attendees as it focuses on the engineering aspects of developing and deploying applications that use streaming data to create new knowledge.

© Springer Nature Switzerland AG 2018
C. d'Amato and M. Theobald (Eds.): Reasoning Web 2018, LNCS 11078, pp. 223–226, 2018.
https://doi.org/10.1007/978-3-030-00338-8_8

This tutorial aims at introducing different existing approaches for querying and reasoning over data streams and providing guidelines to develop and deploy Stream Reasoning applications. In particular, the tutorial offers to the audience: (i) an overview of the use cases and the scenarios where Stream Reasoning can be used (with the advantages it brings); (ii) an overview of the current state of the art in this emerging area, with techniques and tools developed by several research groups (including but not limited to presenters' ones); (iii) a focus on a subset of the technologies to perform complex reasoning over dynamic data.

Related Events: The tutorial follows from the Stream Reasoning for Linked Data (SR4LD) tutorial series, successfully held at ESWC 2011, SemTech 2011, ISWC 2013, ISWC 2014, ISWC 2015; the RDF Stream Processing (RSP) tutorial series at ESWC 2014 and ISWC 2016; the Stream Reasoning: Managing Velocity and Variety in Big Data tutorial at DEBS 2016; and the tutorial on How to Build a Stream Reasoning Application co-located with ISWC 2017.

Technologies are now mature and reliable enough to make a step further and build a tutorial with a stronger focus on hands-on sessions. Therefore, this tutorial not only surveys existing research outcomes and introduce existing tools, but it also presents a relevant Web-centric use-case and guide the participants through application development of a Web stream processing application.

Audience: The tutorial targets researchers and practitioners interested in approaching the topic of web stream processing (both querying and reasoning) and who want to understand the current state-of-the-art as well as the future directions. The technologies and topics on this tutorial are relevant for people from IoT and sensor communities, as well as social media, pervasive health, oil industry, etc., who produce massive amounts of streaming data.

2   Presenters

Emanuele Della Valle[1] holds a PhD in Computer Science from the Vrije Universiteit Amsterdam and a Master degree in Computer Science and Engineering from Politecnico di Milano. He is assistant professor at the Department of Electronics, Information and Bioengineering of the Politecnico di Milano.

In more than 15 years of research, his research interests covered Big Data, Stream Processing, Semantic technologies, Data Science, Web Information Retrieval, and Service Oriented Architectures. He started the Stream Reasoning research field [2] by focusing the research community's interest on this theme through a number of personal initiatives (workshops, projects, journal papers) that contributed to establishing Stream Reasoning as a recognized research and industrial sector. Stream Reasoning is positioned at the intersection between Stream Processing and Artificial Intelligence. In the Big Data era, it is a method to tame simultaneously the velocity (analyzing data streams to enable real-time decisions) and variety (integrating heterogeneous data) dimensions of Big Data. Starting from 2011, he organized more than 15 tutorials and workshops on

---

[1] http://emanueledellavalle.org/.

Stream Reasoning[2]. His work on Stream Reasoning was applied in analyzing Social Media, Mobile Telecom and IoT data streams in collaboration with Telecom Italia, IBM, Siemens, Oracle, Indra, and Statoil. In 2015, he started up a company (Fluxedo) to commercialize the open source results of Stream Reasoning research.

3   Tutorial Structure

1. **Introduction to WEB Stream Processing**
   1.1 From challenges to opportunities.
   1.2 Paradigmatic shit to continuous semantics.
   1.3 The limited ability of existing Stream Processing systems to address variety.
   1.4 The Stream Reasoning research question.
   1.5 Existing Stream Reasoning systems (quick introduction and high-level comparison).
   1.6 Anatomy of a Web Stream Processing Application.
   1.7 Introduction of the tutorial running example.
2. **Publishing and Describing Streams on the Web**
   2.1 Streams on the Web: creation and publications challenges.
   2.2 VoCaLS: a vocabulary to describe RDF Streams on the WEB [5,6]
   2.3 VoCaLSing existing data streams to catalog them[3].
   2.4 TripleWave: a framework to publish RDF streams on the Web [7][4].
   2.5 Hands-on VoCaLS and TripleWave.
3. **Processing Web Streams**
   3.1 Querying and Reasoning on Web Streams: RDF Stream Processing (RSP) languages [8].
   3.2 Overview and comparison RSP query languages: C-SPARQL [9], CQELS-QL [10], $SPARQL_{stream}$ [11], RSP-QL [8].
   3.3 Overview and comparison RSP engines: C-SPARQL E.[5], CQELS[6], Yasper[7];
   3.4 Modeling an RSP Service with VoCaLS.
   3.5 Hands-on RSP engine.
4. **Building a Web Streams Application**
   4.1 Introduction of the RSP Services [12].
   4.2 Designing the Application.
   4.3 Using Jupyter notebooks and the RSP-Kernel[8] to implement the Application.
   4.4 Hands on session.
5. **Wrap-up and discussion**
   5.1 On-going research trends, real-world deployments.
   5.2 Open problems and future directions.

---

[2] http://streamreasoning.org/events.

[3] https://w3id.org/rsp/vocals.

[4] https://github.com/streamreasoning/TripleWave.

[5] https://github.com/streamreasoning/CSPARQL-engine.

[6] https://github.com/danhlephuoc/cqels.

[7] https://github.com/streamreasoning/yasper/.

[8] https://github.com/riccardotommasini/rsp-kernel.

References

1. Dell'Aglio, D., Della Valle, E., van Harmelen, F., Bernstein, A.: Stream reasoning: a survey and outlook. Data Sci. 1(1–2), 59–83 (2017)
2. Della Valle, E., Ceri, S., van Harmelen, F., Fensel, D.: It's a streaming world! reasoning upon rapidly changing information. IEEE Intell. Syst. 24(6), 83–89 (2009)
3. Margara, A., Urbani, J., van Harmelen, F., Bal, H.E.: Streaming the web: reasoning over dynamic data. J. Web Sem. 25, 24–44 (2014)
4. Della Valle, E., Dell'Aglio, D., Margara, A.: Taming velocity and variety simultaneously in big data with stream reasoning: tutorial. In: DEBS, pp. 394–401. ACM (2016)
5. Sedira, Y.A., Tommasini, R., Della Valle, E.: Towards VoIS: a vocabulary of interlinked streams. In: ISWC DeSemWeb (2017)
6. Dell'Aglio, D., Le Phuoc, D., Le-Tuan, A., Ali, M.I., Calbimonte, J.-P.: On a web of data streams. In: ISWC DeSemWeb (2017)
7. Mauri, A., et al.: Triplewave: spreading RDF streams on the web. In: The Semantic Web - ISWC 2016 - 15th International Semantic Web Conference, Kobe, Japan, 17–21 October 2016 (2016)
8. Dell'Aglio, D., Della Valle, E., Calbimonte, J., Corcho, Ó.: RSP-QL semantics: a unifying query model to explain heterogeneity of rdf stream processing systems. Int. J. Semant. Web Inf. Syst. 10(4), 17–44 (2014)
9. Barbieri, D.F., Braga, D., Ceri, S., Della Valle, E., Grossniklaus, M.: C-SPARQL: a continuous query language for RDF data streams. Int. J. Semant. Comput. 4(1), 3–25 (2010)
10. Le-Phuoc, D., Dao-Tran, M., Xavier Parreira, J., Hauswirth, M.: A native and adaptive approach for unified processing of linked streams and linked data. In: Aroyo, L. (ed.) ISWC 2011. LNCS, vol. 7031, pp. 370–388. Springer, Heidelberg (2011). https://doi.org/10.1007/978-3-642-25073-6_24
11. Calbimonte, J., Corcho, Ó., Gray, A.J.G.: Enabling ontology-based access to streaming data sources. In: The Semantic Web - ISWC 2010 - 9th International Semantic Web Conference, ISWC, Revised Selected Papers, Part I, Shanghai, China, 7–11 November 2010, pp. 96–111 (2010)
12. Balduini, M., Della Valle, E.: A restful interface for RDF stream processors. In: Proceedings of the ISWC 2013 Posters & Demonstrations Track, Sydney, Australia, 23 October 2013, pp. 209–212 (2013)

Reasoning at Scale (Tutorial)

Jacopo Urbani[✉]

Vrije Universiteit Amsterdam, Amsterdam, The Netherlands
jacopo@cs.vu.nl

Abstract. This tutorial gives an overview of current methods for performing reasoning on very large knowledge bases. The first part of the lectures is dedicated to an introduction of the problem and of related technologies. Then, the tutorial continues discussing the state-of-the-art for reasoning on very large inputs with particular emphasis on the strengths and weaknesses of current approaches. Finally, the tutorial concludes with an outline of some of the most important research directions in this field.

1 Introduction

Broadly speaking, reasoning can be defined as the task of deriving non-trivial implicit knowledge with a number of logic-based steps. Performing efficient reasoning is an important problem in the area of Artificial Intelligence since the discovery of new knowledge from existing knowledge bases is one of the most important features that intelligent agents should encode.

If we compare reasoning to other related technologies for knowledge discovery (e.g., methods which perform inference using correlation-based embeddings – see [8] for a survey), then we observe that reasoning has the distinctive advantage of providing a clear and unambiguous explanation of the obtained results. Moreover, it is easier to verify that the output of reasoning is correct, and this can be crucial in contexts where even small errors cannot be tolerated (e.g., consider the diagnosis of some diseases, or autonomous driving). Finally, reasoning is useful also for "non-AI" tasks like data integration. To illustrate this last case, consider a fictional company where there are two divisions which are both working on similar products but use a different terminology. In this context, reasoning can be applied to find some links between the knowledge produced by these two divisions to maximize the transfer of knowledge.

Unfortunately, the size of some knowledge bases is particularly challenging to handle for reasoners, also because the input might contain mistakes which lead to inconsistencies. The inability of handling properly large inputs is an important limitation which precludes the application of reasoning to many realistic use cases, like the Semantic Web. This limitation motivated a significant amount of research to develop methods for performing reasoning at scale, especially focusing on Semantic Web data as a realistic "large-scale" use case.

The goal of this tutorial is to (1) introduce the problem of scalability for reasoning; (2) discuss the current state-of-the-art methods and (3) point to existing

C. d'Amato and M. Theobald (Eds.): Reasoning Web 2018, LNCS 11078, pp. 227–235, 2018.
https://doi.org/10.1007/978-3-030-00338-8_9

limitations and other future research challenges. To facilitate the comprehension of the lectures, this document contains a brief explanation of some of the most important concepts covered in the tutorial and links to a number of additional pointers to aid the explanation during the lectures. We assume that the participants have a basic familiarity with standard concepts in Artificial Intelligence and the Semantic Web such as RDF [6], SPARQL [9], or Datalog [1]. If this is not the case, then textbooks like [1,2] offer a first introduction to them.

2   The Problem of Scale

We focus on reasoning that can be performed through an exhaustive execution of rules. Intuitively, rules can be seen as *if-then* constructs which allow the deduction of some conclusions given a set of premises. More formally, a rule is a sentence in first-order logic of the form $\alpha \rightarrow \beta$ where α is the premise (or *body*) and β is the conclusion (or *head*). In our context, α and β are expressions about the truth values of statements that capture the knowledge in the KB. For instance, let us assume that the input knowledge base contains statements about the living location of people and their spouses. These can be expressed as a number of facts $livesIn(a, l), livesIn(b, m), marriedWith(a, c)$ where a, b, c are people, l, m are locations, and $lives, marriedWith$ are two predicates that indicate the living location and respective spouse. Let A, B, L be some generic variables. Then, the rule

$$livesIn(A, L) \wedge married(A, B) \rightarrow livesIn(B, L) \tag{1}$$

captures the implication that married people live in the same location. In this case, α is a conjunction of two atoms while β is a single atom. Notice that rules can be much more complex than (1). For instance, they could contain negated atoms or disjunctions. In this tutorial, we restrict ourselves to consider rules where the body and head are either a single or a conjunction of positive atoms. We say that a rule is *safe* if every variable that appears in the body also appears in the head. Otherwise, we assume that every variable in the head that does not appear in the body is existentially quantified and refer to that rule as an *existential* rule. Existential rules are very important in the Semantic Web because it is well-known that knowledge bases are highly incomplete. Therefore, in some cases it is necessary to reason also about individuals that cannot be identified. For instance, the rule

$$marriedTo(A, B) \rightarrow weddingLocation(A, L) \tag{2}$$

states that if a person A is married, then she/he did so in a certain location L. However, the binary relation $marriedTo$ does not allow us to identify the precise location. We can only say that such location must exist.

Existential rules are more problematic than safe rules because their execution introduces new individuals that represent unknown values. This, in turns, can lead to an infinite computation. While several restrictions can be imposed to ensure termination, in the general case it is not possible to guarantee that the process will always terminate.

To understand the challenges behind the rule execution, we first need to properly define what we mean with "executing a rule". For simplicity, we restrict to the case of safe rules. Let P be the set of rules that we want to execute and $r : B_1, \ldots, B_n \rightarrow H$ be a generic rule in P. Then, let σ be a mapping from variables to constants and use it as a postfix operators for translating atoms into facts (i.e., atoms without variables). For instance, if $X = married(A, B)$ and $\sigma = \{A \rightarrow john, B \rightarrow peter\}$ then $X\sigma = married(john, peter)$. Then, we define $r(I) = \{H\sigma \mid B_1\sigma, \ldots, B_n\sigma \in I\}$ as the set of derivations that rule r can produce from a set of atoms I. Similarly, $P(I) = \cup_{r \in P} r(I)$ is the set of derivations produced by all rules on I. Clearly, rules can be executed exhaustively. To this end, let $P^0(I) = I$ and $P^{i+1}(I) = P(P^i(I)) \cup P^i(I)$ for $i \geq 0$. The set $P^\infty(I)$ is called the *materialization* of I using P and it is typically what we want to compute.

We identify five classes of challenges for computing $P^\infty(I)$:

- **C1:** The input set I might be too large to be stored and accessed efficiently with the given hardware. For instance, consider that the linked data cloud is estimated to contain about 30 billion statements. If we make the assumptions that each statement can be stored in about 24 bytes (that is, each term and predicate is represented by a 64bit number), then storing this large collection would take about 686 GB of memory.
- **C2:** Similarly as before, the set $P^\infty(I)$ might be too large to be stored with the given hardware (this can occur where the rules produce an explosion of the derivations).
- **C3:** The execution of some rule r on a given I might be particularly challenging because the generation of all possible σ requires a join between atoms in I. It is well-known that the order of $B_1, \ldots, B_n$ can determine which join algorithm can be used and this can have a significant impact on the performance. Unfortunately, it is not always easy to determine the best ordering beforehand.
- **C4:** There can be two rules which can derive a significant number of equivalent derivations, or the same rule can derive the same conclusions if applied on the augmented versions of the same database. The duplicate derivations need to be detected and removed.
- **C5:** There can be rulesets and/or particular input databases that trigger a very large number of rule executions before the reasoner reaches the point when $P^{i+1}(I) = P^i(I)$. As we mentioned before, if the rules are not safe then this can even degenerate in an infinite computation.

We say that a reasoning approach (or technique, or system) is *scalable* if it can deal efficiently with at least one of these challenges.

3   Reasoning at Scale

The scalability of reasoning can be improved on two dimensions: Either by utilizing more efficiently the computing resources of a single machine (for instance by compressing the data, or by exploiting modern multicore architectures) or by distributing the computation on different machines. The first approach is eventually limited by the physical capabilities of the underlying hardware while the second one is limited by the network bandwidth and potential bottlenecks synchronization mechanisms. In the tutorial, we cover both types of approaches. First, we discuss approaches to distribute the computation of the full materialization since these were the first scalable approaches proposed in the literature. Then, we describe the centralized approaches, which are more recent. Finally, we discuss the execution of query-driven reasoning and of existential rules.

3.1 Distributed Materialization

We can distribute the computation of rule-based reasoning following two approaches: First, we can replicate the input data on each machine and execute each rule concurrently. We refer to this form of distribution as *inter-rule* distribution since multiple rules are executed concurrently. Second, we can partition the data and execute the rules only on the data that is locally available. We refer to this approach as *intra-rule* distribution since one rule is processed by multiple machines. With both types of distribution, there can be cases when the machines do not need to exchange information during the computation. This represents the best case since we can obtain an ideal speed up. Unfortunately, most of the times the machines need to communicate with each other. In this case, the communication can quickly become a performance bottleneck that limits the performance irrespectively of the number of machines.

In the context of reasoning on Semantic Web data, the *intra-rule* parallelism has returned encouraging results. More in particular, MapReduce-based reasoners like WebPIE [12] were able to compute the closure of large databases with billions of facts using a moderate number of machines. This performance was obtained by partitioning the input and output of reasoning on different machines, and by executing the rules on multiple partitions at the same time. Moreover, several optimizations are introduced to either avoid the generation of duplicates. We illustrate one example of such optimization below. Additional optimizations are discussed during the tutorial.

Assume we want to execute the following two rules:

$$T(A, P, B), T(P, \mathtt{rdfs:domain}, C) \rightarrow T(A, \mathtt{rdf:type}, C) \qquad (3)$$

$$T(A, P, B), T(P, \mathtt{rdfs:range}, C) \rightarrow T(B, \mathtt{rdf:type}, C) \qquad (4)$$

where T is a generic predicate, A, B, C, P are variables, and $\mathtt{rdfs:domain}$, $\mathtt{rdfs:range}$, and $\mathtt{rdf:type}$ are standard constants from the RDF and RDFS vocabularies. Intuitively, these two rules capture the domain and range of properties and use them to infer the classes of the arguments of these relations. Unfortunately, both rules can derive the same conclusion.

For instance, let us assume that we have two properties, $\mathtt{playsFor}$ and $\mathtt{taughtBy}$ which have $\mathtt{Person}$ as domain and range respectively. In this case, it is easy to see that both rules will produce the same derivation if there is an entity with both relations. To avoid this case, we can write a simple MapReduce program where the *map* function outputs, for each input triple, two tuples which have as key either the subject or object of the input triple. Then, the framework will group all intermediate pairs by key, which in our case is a single entity. Each group becomes the input of the *function* which is applied concurrently on different machines. This function can be implemented so that it processes all the triples in the group and whenever they instantiate the body of either (3) or (4) then a derivation is returned. These derivations will have the group key as subject, following the logic of the rules. Therefore, since groups are partitioned by the key, it is not possible that two different groups will return the same derivation, avoiding thus the possibility of producing multiple copies concurrently.

A limitation of WebPIE is that it can execute only a specific set of rules. We are not aware of extensions to these methods which allow the execution of generic rules and can scale to similar inputs. This feature is supported by centralized methods, which are the topics of the following subsection.

3.2 Centralized Materialization

More recently, two scalable approaches were proposed for performing reasoning using shared-memory architectures. These approaches are interesting because shared-memory architectures are typically cheaper than MapReduce clusters and are available in more contexts. These approaches differ from each other because they improve the scalability in different ways. The first approach, which is implemented in the system RDFOx [7], focuses on exploiting the multicore architecture avoiding as much as possible any locking mechanism that impedes parallelism. The second approach, implemented in the system VLog [11], focuses instead on improving the performance by storing the derivations using a columnar layout rather than a more conventional row-by-row layout.

The strategy adopted by the system RDFox is to store the input and the derived facts into a number of highly-optimized hash tables designed which are almost-lock-free (i.e., they are data structures which can be updated concurrently with little or no delay due to locking). Then, it stores the facts that might lead to new derivations in a centralized queue and let concurrent threads attempt at producing new derivations using its content. New derivations are continuously added to the hash tables and to the queue for further consideration until fixpoint. By doing so, the system can effectively parallelize the execution of the rules and is able to exploit all available cores to the fullest.

VLog is another reasoner for shared-memory architectures which improves the scalability by storing the derivations using a columnar layout. For instance, consider a simple binary relation R with the following facts: $R(a_1, b_1)$, $R(a_1, b_2)$, $R(a_1, b_3)$, $R(a_2, b_1)$. In this case, the columnar layout would store the content of R using two columns $c_1 = \langle a_1, a_1, a_1, a_2 \rangle$ and $c_2 = \langle b_1, b_2, b_3, b_1 \rangle$. This storage layout is useful because it allows the implementation of some compression techniques to reduce the storage of the derivations. For example, if we look at the content of c_1, then we can observe that the first three elements can be compressed by simply appending after the first element a_1 a special sequence that indicates that the value should be repeated three times. This type of well-known compression, which is called run-length encoding (RLE), effectively reduces the used space and enables the storage of a larger number of derivations. The usage of columns is also interesting because columns can be reused in some specific cases. This further reduces the space and enhances the scalability.

In practice, both RDFOx and VLog can handle inputs with billions of statements without requiring large resources. For instance, VLog is able to compute reasoning over the entire DBPedia using a single laptop. Moreover, they have the additional advantage that the derived data is already indexed, and this enables an efficient querying of the inferred knowledge without expensive loading operations.

3.3   Query-Driven Reasoning

There are cases when performing a full materialization is either undesirable or impossible. For instance, consider the case where the knowledge base contains some errors and the full materialization would result in a huge number of non-sensical derivations.

In such a case, a partial materialization might allow us to retrieve at least a subset of all derivations. To this end, there are some well-known techniques which we can use to rewrite the original program in such a way that it would only derive facts that are relevant for answering a given query. More formally, let us assume that the user is only interested in answering a specific query Q, which is represented by a single atom, over $P^\infty(I)$. In this case, the objective of the system is to compute only the derivations that are strictly necessary to retrieve all answers for Q.

The most famous technique for performing this type of rewriting is *Magic Sets* [3]. In essence, this technique introduces a number of additional predicates, called magic predicates, to produce only those derivations that might lead to compute potential answers. It is possible to replicate the behavior of magic sets by using auxiliary data structures without any program rewriting. This operation is performed, for instance, by the algorithm QSQR [1], which rewrites an initial query into a number of subqueries, and caches the results of previously-executed queries to ensure termination.

Currently, query-driven reasoning is supported by state-of-the-art reasoners like RDFox or VLog. One additional system that performs this type of reasoning in a distributed setting is QueryPIE [14]. This system implements the QSQR algorithm introducing a couple of novelties that improve the scalability. The first novelty is that it pre-computes the answers to some specific queries, namely the ones that describe the schema of the data. This pre-computation is important because the pre-computed queries are relevant for answering multiple queries. Second, the system "weakens" the rule execution by allowing the execution of the same query multiple times. This operation results in some redundant computation but it has the advantage that it reduces the number of synchronization barriers and thus it can be executed on multiple machines.

3.4 Existential Rules

It is interesting to drop the requirement that every variable in the head must appear in the body of the rule in order to reason about unknown individuals. For instance, consider the rule

$$isPerson(X) \rightarrow isParentOf(X,Y) \tag{5}$$

which states that every person should have at least one parent. In this case, we do not know who is the parent, but we know that such parent must exist. This type of rules, which we call existential rules, is more problematic to execute because they require the introduction of fresh individuals and it is easy to show that there are cases when this operation might lead to a non-termination of the rule application. While in some cases there is no alternative than stopping the process without any guarantee of completeness, in some cases we can perform some checks to verify that the process will terminate (for instance, some of these checks are presented in [5]). Some recent work has shown that in practice most of the real-world ontologies satisfy these checks. Moreover, we can restrict the number of derivations by allowing the rules to produce a new derivation only if the head of the rule was not previously instantiated with the non-existentially quantified variables. In some cases, this constraint allows the computation to terminate.

Regardless of the restrictions applied during the computation of these rules (and the problem of non-termination), the management of the fresh individuals makes it is challenging to execute existential rules on large inputs. Recently, a new benchmark was introduced to facilitate the development and comparison of systems for the execution of existential rules [4]. Using the test cases provided by this benchmark, it emerged that the columnar architecture of VLog is well-suited also for the execution of these rules [13]. This means that it is possible to perform scalable reasoning also with existential rules, and this opens the door for the application of reasoning to a much larger number of use cases.

4 Discussion

In Sect. 2 we indicated five classes of challenges that reasoners must tackle in order to be able to process large inputs. It is interesting to relate these challenges to the systems covered in the previous sections to understand how they effectively address them.

Challenge C1. In general, all systems attempt at compressing the input so that more data fits in one machine. The most common approach consists of dictionary-encoding the data so that constants and predicates are represented by (shorter) numerical IDs. Smarter compression techniques (e.g., [10]) can be used to further reduce the occupied space. Distributed approaches like WebPIE or QueryPIE further address this challenge by partitioning the input on multiple machines.

Challenge C2. The second challenge is concerned with the storage of the derivations. This challenge is different than the previous one because some compression techniques that can be applied to the input cannot be applied to the set of derived data since this is continuously augmented. To the best of our knowledge, only VLog addresses this challenge by adopting the columnar storage layout.

Challenge C3. The problem of detecting the optimal join order has been studied extensively in database literature. Current reasoners address this issue in various ways. WebPIE uses different join algorithms depending on the input rule while VLog chooses the best ordering using cardinality estimation, and employs merge join whenever the data is properly indexed. In contrast, RDFox relies on hash-based joins leveraging the highly optimized hash-based indices. Despite these efforts, finding the best join ordering remains a challenging problem and this is still one of the major causes that can slow down the reasoning process.

Challenge C4. This challenge is addressed in the WebPIE system with the introduction of special algorithms which "group together" the execution of multiple rules that might produce the same derivations. In this way, duplicates can be removed locally with cheaper operations. Moreover, VLog partly addresses this issue by performing backward reasoning in order to determine whether the execution of a given rule would return some duplicates. Static analysis on the program might further help to unveil some potential rule execution orders which might lead to duplicates, but it is essential that the system maintains some sort of provenance of the derivations. Currently, only VLog supports this feature.

Challenge C5. This challenge is intrinsic to the task of reasoning and it is easy to construct examples which require a huge amount of computation and cannot be easily parallelized. In these cases, one solution consists of adopting an anytime behavior and return new derivations as soon as they are derived. Query-driven approaches mitigate this problem by restricting reasoning to given queries. In this way, in some cases reasoning can still terminate quickly. In general, none of the other reasoners can circumvent this problem without dropping completeness. It is interesting to explore whether predictive techniques can determine with some degree of confidence what is the likelihood that the reasoner has computed a certain ratio of all derivations. This would allow a premature stop of the process with some indication of how much is still missed. To the best of our knowledge, this is an open problem which, if solved, might significantly improve the scalability of existing systems.

References

1. Abiteboul, S., Hull, R., Vianu, V.: Foundations of databases, vol. 8. Addison-Wesley Reading, Boston (1995)
2. Antoniou, G., Van Harmelen, F.: A Semantic Web Primer. MIT press, Cambridge (2004)
3. Bancilhon, F., Maier, D., Sagiv, Y., Ullman, J.D.: Magic sets and other strange ways to implement logic programs. In: Proceedings of the Fifth ACM SIGACT-SIGMOD Symposium on Principles of Database Systems, pp. 1–15 (1985)

4. Benedikt, M., et al.: Benchmarking the chase. In: Proceedings of the 36th ACM SIGMOD-SIGACT-SIGAI Symposium on Principles of Database Systems, pp. 37–52. ACM (2017)
5. Grau, B.C., et al.: Acyclicity notions for existential rules and their application to query answering in ontologies. J. Artif. Intell. Res. **47**, 741–808 (2013)
6. Hayes, P.: RDF Semantics. W3C Recommendation (2004)
7. Nenov, Y., Piro, R., Motik, B., Horrocks, I., Wu, Z., Banerjee, J.: RDFox: A highly-scalable RDF store. In: Arenas, M. (ed.) ISWC 2015. LNCS, vol. 9367, pp. 3–20. Springer, Cham (2015). https://doi.org/10.1007/978-3-319-25010-6_1
8. Nickel, M., Murphy, K., Tresp, V., Gabrilovich, E.: A review of relational machine learning for knowledge graphs. Proc. IEEE **104**(1), 11–33 (2016)
9. Prud'hommeaux, E., Seaborne, A.: SPARQL Query Language for RDF. W3C Recommendation (2008)
10. Urbani, J., Dutta, S., Gurajada, S., Weikum, G.: KOGNAC: efficient encoding of large knowledge graphs. In: Proceedings of IJCAI, pp. 3896–3902 (2016)
11. Urbani, J., Jacobs, C., Krötzsch, M.: Column-oriented datalog materialization for large knowledge graphs. In: Proceedings of AAAI (2016)
12. Urbani, J., Kotoulas, S., Maassen, J., Van Harmelen, F., Bal, H.: WebPIE: a web-scale parallel inference engine using MapReduce. Web Semant.: Sci. Serv. Agents World Wide Web **10**, 59–75 (2012)
13. Urbani, J., Krötzsch, M., Jacobs, C., Dragoste, I., Carral, D.: Efficient model construction for horn logic with VLog. In: Galmiche, D., Schulz, S., Sebastiani, R. (eds.) IJCAR 2018. LNCS, vol. 10900, pp. 680–688. Springer, Cham (2018). https://doi.org/10.1007/978-3-319-94205-6_44
14. Urbani, J., Piro, R., van Harmelen, F., Bal, H.: Hybrid reasoning on OWL RL. Semant. Web **5**(6), 423–447 (2014)

Author Index

Printed in the United States
By Bookmasters